DESIGN OF
RARE-EARTH PERMANENT MAGNETS
AND
MAGNETIC CIRCUITS

BY

HERBERT A. LEUPOLD

AND

ERNEST POTENZIANI II

Wexford Press
2008

Table of Contents

List of Figures

List of Tables

1. Introduction

The advent of rare-earth permanent magnets (REPMs) has brought the realization of novel magnetic structures that are not practicable otherwise.[1,2] So different are these remarkable materials from the earlier magnets that conventional design wisdom is inadequate to fully exploit their unique characteristics. Indeed, the conventional wisdom can lead to error or to the employment of cumbersome procedures that are quite unnecessary for REPMs. The salutary characteristics stem from two basic attributes of the rare earth materials: 1) large intrinsic moments per unit volume; 2) extraordinarily high resistance to demagnetization by external or internal demagnetization fields.

The magnetization offers high flux density; the coercivity enables the magnets to maintain this flux density in the face of very high demagnetizing fields engendered by flat aspect ratios. Thus, REPMs can be fashioned, with impunity, into shapes that would cause demagnetization of alnico materials. This point is illustrated in Fig. 1, which shows the second quadrant of the hysteresis loop of a typical REPM. Note that the constancy of magnetization in reverse fields of more than 1.3 Tesla (T) results in a linear B versus $\mu_0 H$ curve of slope one throughout the entire quadrant. This linearity of the magnetization curve makes possible magnetic structures in which the magnet is exposed to demagnetizing fields in excess of the remanence. The importance of this property to design simplification will become obvious from the following discussions in which $\mu_0 H$ and $\mu_0 M$ will be written B_0 and J respectively for convenience in curve plotting and calculations. This procedure has the advantage that in the relationship between field, flux density and magnetization, $B = B_0 + J$, all three of the principal quantities involved are of the same kind and measured in T.

Magnetic circuit design with the older magnetic materials, such as alnico, is largely an intuitive hit-or-miss affair, as is clear from the B vs B_0 curve for alnico in Fig. 1. Flux densities are low except for geometries with the very highest B to B_0 ratios, that is, those with the very lowest demagnetizing factors. Because the state of magnetization of each magnet depends upon the geometry of the structure in which it is placed, exact analytic solution for all but the simplest configurations is very difficult or impossible. Hence, much reliance is placed on numerous approximations, actual physical models, electric analogues, or rules of thumb.

Because of the analytical simplification afforded by the magnetic rigidity of the REPMs, the range of applicability of traditional approaches to magnetic design has been greatly extended. These approaches fall into four broad types: 1) analogy of magnetic configurations to electrical circuits; 2) analytical solutions through Maxwell's equations; 3) reduction of permanent magnet arrays to distributions of equivalent pole densities or current sheets and the insertion of those distributions into Coulomb's law or the Biot-Savert law; and 4) brute-force computer solution of a tentative configuration, the plausibility of which is previously established by one of the other approaches, usually the first. In this work we will discuss these approaches and employ them to solve illustrative problems.

2. Magnetic Circuit Design

2.1 Magnetic Analogue to Ohms Law

In the past, elementary courses in general physics often included a brief treatment of magnetic

1

circuits that featured a magnetic analogue of Ohm's law. In this scheme, the role of electric current I is played by the magnetic flux Φ, that of the electromotive force V by the magnetomotive force F, the electric conductance G by the magnetic permeance P, and the resistance R by the reluctance R, so that the magnetic Ohm's law reads:

$$\Phi = PF = F/R$$

in respective correspondence to

$$I = GV = V/R \tag{1}$$

Permanent magnets can then be regarded formally as magnetic "batteries," and materials of high permeability, such as soft iron or permalloy, as near-perfect flux conductors, in analogy to the near-perfect current conductor, copper, in electrical circuits. Air gaps and materials of low permeability play the role of magnetic resistors. The analogy is completed by the identification of electrical solenoids wound about permeable circuit members as flux "generators," which can be either ac or dc, depending upon the currents sent through them.

Although this approach, like most analogies, is philosophically and mnemonically gratifying, it was rarely used in practice as outlined, and has now been dropped from the elementary courses. The barrier to more general usefulness was essentially twofold. In the first place, unlike electric currents, magnetic fluxes are not confined to neat, analytically tractable paths like wires, but fill virtually all of space. This difficulty is not too serious, because, in many cases, the space around a magnetic circuit can be divided into flux paths of which the boundaries are planes, cylindrical arcs, or segments of spherical surfaces that emanate normally from the surfaces of the circuit to connect points of different magnetic potential. The permeances of these simplified paths can be calculated approximately through standard formulae, and if the division is made judiciously, surprisingly good approximations to the true fluxes can often be obtained. Figure 2 shows some frequently used flux paths together with the formulae used to calculate their permeances.

The second, and more serious, difficulty that has prevented a simple, direct application is that magnetic materials, such as alnicos, generate no unique magnetomotive force (mmf), because the mmfs depend upon the circuit into which they are inserted. The great value of REPMs in the simplification of circuit design arises because these materials exhibit the same total circuital mmf, regardless of the nature of the circuit in which they find themselves. Although this advantage has been discussed previously,[1-3] it is still not as widely appreciated as are the high energy products and coercivities of the REPMs; consequently, circuits employing these materials are still often analyzed by unnecessarily cumbersome procedures.

To illuminate the origins of these difficulties and to highlight the contrasts between the rare earths and their predecessors, it is useful to consider the derivation of the magnetic Ohm's law. We begin by writing the circuital form of Ampere's law

$$\oint \vec{H} \cdot d\vec{s} = \pm I \tag{2}$$

2

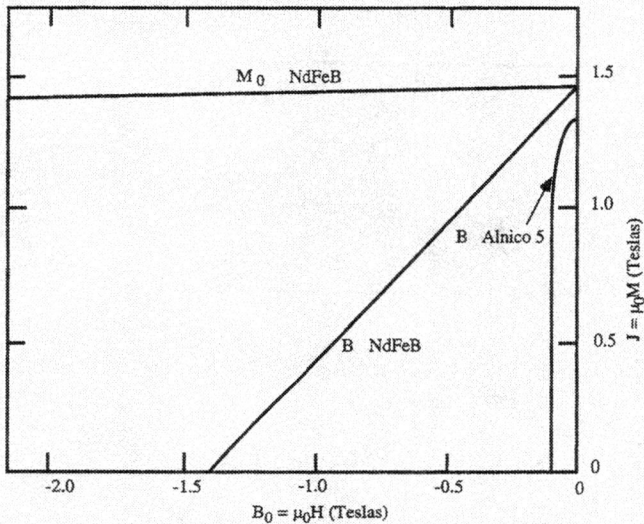

Figure 1. Demagnetization curves for NdFeB and alnico 5 magnets.

where the magnetic field H is in amps/meter, the current I, linked by the loop integral, is in amperes, and the sign depends upon whether the current direction in the coils produces an mmf aiding or opposing that of the magnet. Assuming H_m is the average field over the length of the path in the magnet L_m, we can write

$$H_m L_m + \int_A^B \bar{H}_L \cdot d\bar{s} = \pm I \tag{3}$$

where A and B are the ends of the magnet, the line integral is along a flux line outside of the magnet, and H_L is the field along that line. Outside the magnet, $\mu_L H_L = B_L$, therefore

$$H_m L_m + \int_A^B \frac{\bar{B}_L \cdot d\bar{s}}{\mu_L} = \pm I \tag{4}$$

but

$$B_L = \Delta\Phi/\Delta A$$

where ΔA is a differential element of area normal to a flux line of differential length ds and $\Delta\Phi$ is the element of flux passing through it. Consequently,

3

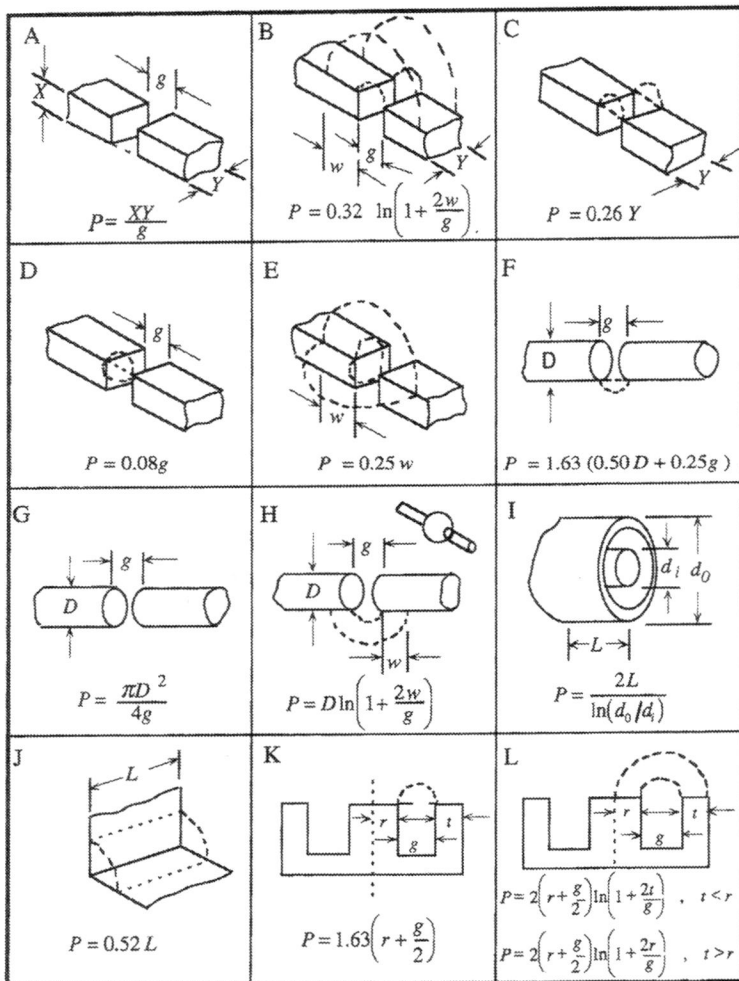

A

$$P = \frac{XY}{g}$$

B

$$P = 0.32 \ \ln\left(1 + \frac{2w}{g}\right)$$

C

$$P = 0.26 \, Y$$

D

$$P = 0.08g$$

E

$$P = 0.25 \, w$$

F

$$P = 1.63 \, (0.50 \, D + 0.25g)$$

G

$$P = \frac{\pi D^2}{4g}$$

H

$$P = D \ln\left(1 + \frac{2w}{g}\right)$$

I

$$P = \frac{2L}{\ln(d_0 / d_i)}$$

J

$$P = 0.52 \, L$$

K

$$P = 1.63\left(r + \frac{g}{2}\right)$$

L

$$P = 2\left(r + \frac{g}{2}\right)\ln\left(1 + \frac{2t}{g}\right) \ , \ \ t < r$$

$$P = 2\left(r + \frac{g}{2}\right)\ln\left(1 + \frac{2r}{g}\right) \ , \ \ t > r$$

Figure 2. Standard permeances

$$\pm I = H_m L_m + \int_A^B \frac{\Delta \Phi}{\mu_L \Delta A} ds = H_m L_m + \Delta \Phi \int_A^B \frac{ds}{\mu_L \Delta A} = 1 \tag{5}$$

Of course, the area ΔA and permeability μ_L associated with flux $\Delta \Phi$ may vary with progression along the flux line, that is, they are functions of s. We recognize that the expression in the integral is just the differential reluctance

$$dR = \frac{ds}{\mu_L \Delta A}$$

so that

$$\int \frac{ds}{\mu_L \Delta A} = \Delta R_t^E = \frac{1}{\Delta P_t^E} \tag{6}$$

where ΔP_t^E is the permeance of the tube enclosing flux $\Delta \Phi$ exclusive of the permeance of its path within the magnet. Upon integration over s, Equation (5) becomes:

$$\frac{\Delta \Phi}{\Delta P_t^E} = \pm I - H_m L_m \tag{7}$$

Since these tubes are all in parallel and fill all of space, their summation and that of their fluxes yields

$$\Phi_m / P_t^E = \pm I - H_m L_m \tag{8}$$

where Φ_m is the total flux emanating from the magnet; therefore, using,

$$\Phi_m = B_m A_m \tag{9}$$

where B_m is the average flux density within the magnet, we obtain

$$B_m = P_t^E [\pm I - H_m L_m] / A_m \tag{10}$$

For NdFeB the demagnetization curve is reversible, linear, and of the form

$$B_m = \mu_r H_m + B_r \tag{11}$$

where μ_r is the recoil permeability and B_r is the remanent flux density, $\mu_0 M_r$. Solving for H_m, we obtain

$$H_m = (B_m - B_r) / \mu_r = \Phi_m / \mu_r A_m - B_r / \mu_r \tag{12}$$

5

Multiplication of (10) by A_m and combination of the results with (12) yields

$$\Phi_m = P_t^E \left[-\Phi_m L_m / \mu_r A_m + B_r L_m / \mu_r \pm I \right] \tag{13}$$

$$\Phi_m = \frac{\left(B_r L_m / \mu_r \pm I \right)}{\left(R_i^E + R_m \right)} \tag{14}$$

This is the magnetic form of Ohm's law where the total flux Φ_m is analogous to the total current and the reluctances $R_i^E = 1/P_t^E$ and $R_m = L_m / \mu_r A_m$ correspond to the total circuit resistance external to the battery and the battery internal resistance, respectively. Corresponding to the battery emf is $L_m B_r / \mu_r$ or $-L_m H_c$ and $\pm I$ is the analogue of the emf from a zero impedance, constant voltage source. This simple and useful formula is applicable to cobalt-rare earth magnets because of the linear relationship between H_m and B_m expressed by Eq. (11) which, in turn, is a consequence of the constancy of magnetization in all parts of the magnet, regardless of stimuli arising from the rest of the circuit or the magnet's own shape. In contrast, alnicos do not have reversible demagnetization curves that are expressible in simple analytical form. Therefore, in dealing with circuits containing such magnets one must proceed less directly and with recourse to B-B_0 graphs such as those shown in Fig. 3.[3] For simplicity, we consider as an example a circuit with no electromagnetic sources of flux. In such a case, Eq. (10) can be written:

$$B_m / H_m = L_m P_t^E / A_m \tag{15}$$

Then, one must calculate P_t^E which, together with the use of magnet length and the cross-sectional area in (15), yields B_m/H_m. Next, a load line with slope $B_m / \mu_0 H_m = B_m/B_{0m}$ is drawn through the origin and its intersection with the second quadrant B-B_0 curve determines B_m. The product $B_m A_m$ then yields Φ_m, which is the total flux output of the magnet; Φ_m is distributed among the various flux paths in the circuit in accordance with Ohm's and Kirchoff's laws.

If the demagnetizing field of the alnico magnet is lessened by an increase in P_t^E resulting from the narrowing of an air gap within the circuit, the operating point of the magnet moves along an approximately linear minor loop to point B in Fig. 3, in accordance with the new value of B_m/B_{0m}. When changes in the circuit are such as to move the operating point from left to right on the minor loop, we can use an expression such as (11) with the intercept of the minor loop with the B axis, B_i in place of B_r, and the slope of the minor loop μ_i / μ_0 in place of μ_r / μ_0. Consequently, one may define a magnet mmf of the same form as that for the rare earth magnets, i.e.,

$$F = B_i L_m / \mu_i \tag{16}$$

However, should the demagnetizing field increase due to a widening of an airgap, temporary removal of the magnet from the circuit, or application of an external demagnetizing field, the operating point would move to the left along the demagnetization curve to C, the base of a new minor loop CD. The magnet would also have a new mmf given by the new line constants B_i' and μ_i' / μ_0

$$F = B_i' L_m / \mu_i' \tag{17}$$

Hence, we see that no unique mmf can be assigned to a conventional permanent magnet, and if the useful magnetic field is to be modulated by variation of gap length or electrically generated mmf's, the magnet mmf will always be that corresponding to the lowest point on the demagnetization curve reached in the course of the modulation cycles. An additional complication in determining the mmf of an alnico magnet is that the lengths L_m appearing in Eqs. (16) and (17) are effective rather than actual lengths. Because of demagnetizing fields, the magnetization tends to be nonuniform, so that different parts of the magnet lie on different load lines, and therefore produce different mmf's and inhomogeneous fields. This causes effective lengths to be shorter than actual ones and all the other germane quantities, such as H_m and B_m, to have only a rough average significance. This problem exists for Rare Earth Magnets as well but to a lesser extent because only H can vary from point to point while for alnicos, both H and M vary. For some configurations, use of geometrical lengths and average field values can cause significant errors, and estimations of effective lengths must be made.[1] Magnetic quantities and their electric analogues are summarized in Table 1.

2.2 Example of Magnetic Gap Field Calculations

In Fig. 4(a) is pictured a simple magnetic circuit with the space around it divided into permeance paths as described in the introduction. The flux and field of interest are those in the magnet gap P_G. To calculate these quantities we must first find the permeances associated with the fourteen external flux paths. All are of standard form and can be calculated by means of formulae found in Fig. 2 and in Refs. 3-6. To aid in the visualization of the forms of these paths, brief descriptions of each are summarized in Table 2. The fourteen permeances are in parallel with each other and in series with the internal permeance P_m of the magnet. (See Fig. 4(b).) Almost rigid magnets ($B_r / B_{0c} = 1.05$) of $B_r = 1.32$ T are employed. From the values of the quantities shown, the total circuit reluctance is calculated to be $R_t = R_m + R_t^g = 3.333 + 0.345 = 3.678$ m^{-1}. The mmf: $F = L_m B_{0c} = L_m B_r / 1.05 = 0.4\,B_r / 1.05 = 0.381\,B_r = 0.503$ T-m. Here, as previously stated, we use B_0 rather than H so that $F = L_m B_C$ is really a quasi-potential measured in Tesla-meters. There is no difficulty involved when this value is used to obtain a flux via the formula $\Phi = P\,F$ because:

$$F(\text{true}) = B_0 L / \mu_0 = F(pseudo)/\mu_0$$

and

$$P(true) = \mu_0 \mu_p A / L = \mu_0\,P(pseudo)$$

where μ_p is the permeability relative to that of empty space.

7

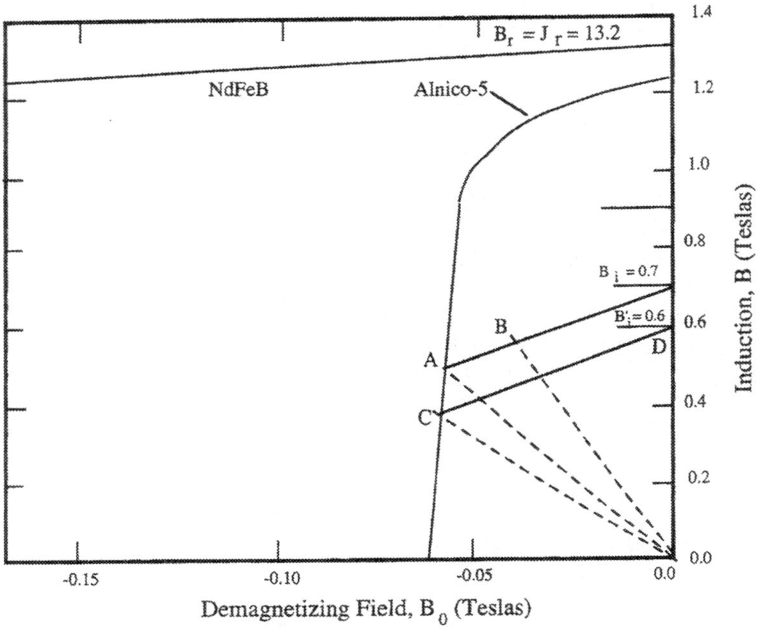

Figure 3. Determination of B_m in an alnico 5 magnet. The load line OA is determined from the total external permeance as described in the text. Bm is the value of B at intersection A. A decrease in P_t^E would lower the operating point to C, the base of a new reversible minor loop CD. The reversible linear demagnetization curve for NdFeB magnets of 14.2 T is shown for comparison.

Now,

$$\Phi = P(true)\, F(true) = \left[F(pseudo)/\mu_0 \right]\left[\mu_0 P(pseudo) \right]$$

$$\Phi = F(pseudo)P(pseudo)$$

so that when $B_0 L$ is used for the potential F, and the relative permeability, $\mu_p A / L$, for P the same results are obtained as from the true values of F and P. In most problems μ_p is either equal to 1.0, as in air gaps and in rigid permanent magnets, or to ∞, as in iron yokes and pole pieces. Rarely do paramagnetic substances occur in design problems but they present no special difficulties, as their μ_p's are between unity and infinity in the permeance formulae. When passive materials such as iron are operated close to or above saturation the approximation $\mu_p = \infty$ does not hold and the method of permeance estimation is not readily practicable.

Therefore we find the total flux emanating from the magnet:

$$\Phi_t = FP_t = (0.503)(0.272) = 0.137$$

8

Table 1.   Magnetic Circuit Quantities and Their Electric Counterparts

MAGNETIC			ELECTRIC		
Magnetomotive Force	$\Delta U, F$	Amperes	Electromotive Force	$\Delta V, \varepsilon$	Volts
Magnetic Potential	U	Amperes	Electric Potential	V	Volts
Magnetic Flux	Φ	Webers	Electric Current	I	Amperes
Permeance	P	Webers/Ampere	Conductance	G	Siemens
Reluctance	R	Ampere/Webers	Resistance	R	Ohms
Magnetic Field	H	Amperes/m	Electric Field	E	Volts/m
Flux Density	B	Teslas	Current Density	J	Coulomb/m^2-sec
Permeability	μ	Tesla-m/Ampere	Conductivity	σ	Siemens/m
Magnetic Field Energy	W	Joules	Electric Power	P	Watts
MMF of a Magnet	$L_m \, H_c$	Amperes	EMF of a Battery	V_b	Volts

9

Since all permeances in P_t^E are in parallel, the gap flux is given by:

$$\Phi_g = P_g \Phi_t / P_t^E = (1)(0.137) / (2.90)$$

and the average field in the gap

$$B_g = \Phi_g/A_g = 0.047/0.2 = 0.236 \text{ T.}$$

If the calculation were to be made for the same configuration with the NdFeB replaced by alnico V, one would proceed as follows:

(1) Calculate the total external permeance P_t^E = 2.901 m and total circuit permeance P_t = 0.2719 m.

(2) Find the ratio B_m/B_{0m} via the formula

$$B_m/B_{0m} = -L_m P_t^E / A_m = -(0.4)(2.901)/(0.12) = -9.67$$

(3) Draw the line B_m/B_{0m} = -9.67 together with the second quadrant of the alnico V demagnetization curve (Fig. 3).

(4) Find the magnet operating point at intersection (A) and the value of B_m at that point, i.e., B_m = 0.514 T, as in Fig. 3.

(5) Calculate total flux $\Phi_t = B_m A_m = 0.062$ Webers.

(6) Find $\Phi_G = P_G \Phi_t / P_t^E = 0.021$ Webers.

(7) Find $B_G = \Phi_G/A_G = 0.107$ T, less than half the field obtained when NdFeB is used.

We see that the first and last two steps in this calculation must also be completed for NdFeB and Φ_t obtained from $\Phi_t = F/R_t$ in lieu of step (5). But the time-consuming steps, (2) to (4), are unnecessary when REPMs are used, and no external aids such as graphs are needed.

Magnetic circuit design by means of the electric circuit analogue, although seemingly crude, yields valuable qualitative guides to design adjustment and, in many cases, good quantitative results as well. The latter are often surprisingly accurate, especially when the calculated quantities of interest involve averages over large areas, as is the case of magnetic flux calculation. If a local field or flux density calculation is made, agreement of calculated values with actual values is not as good. These points are illustrated in Fig. 5.[6] Note the remarkable agreement in geometric detail between the rough permeance calculations and the computer plot of Fig. 5. Point A is the point which, according to the calculations, is

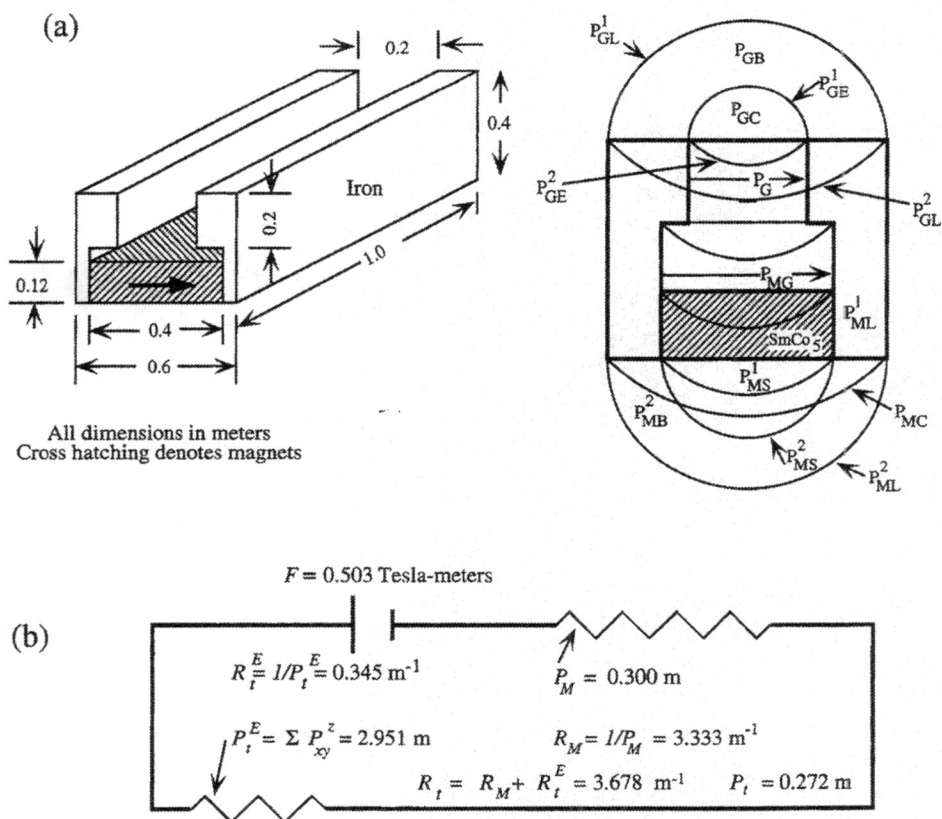

(a)

0.2

0.4

Iron

0.2

1.0

0.12

0.4

0.6

All dimensions in meters
Cross hatching denotes magnets

P_{GL}^1
P_{GB}
P_{GE}^1
P_{GC}
P_{GE}^2
P_G
P_{GL}^2
P_{MG}
P_{ML}^1
SmCo$_5$
P_{MS}^1
P_{MB}^2
P_{MC}
P_{MS}^2
P_{ML}^2

$F = 0.503$ Tesla-meters

(b)

$R_t^E = 1/P_t^E = 0.345 \ m^{-1}$

$P_M = 0.300 \ m$

$P_t^E = \Sigma \ P_{xy}^z = 2.951 \ m$

$R_M = 1/P_M = 3.333 \ m^{-1}$

$R_t = R_M + R_t^E = 3.678 \ m^{-1}$ $P_t = 0.272 \ m$

Figure 4. (a) Division of space around magnetic circuit into approximate flux paths. (b) Schematic of equivalent electric circuit.

the demarcation between the flux lines which go from limb to the iron plate and those which return to the other end of the magnet. Point B is the calculated effective equator of the magnet. Note how it is the approximate symmetry center for the P_o flux lines. Point C is the calculated point that marks the switching of the center post flux lines from the permeable plate to the ring magnet. Point D is the predicted switch-over point from P_r paths to P_b paths. Point E is the demarkation between fluxes emanating from the ring magnet to the iron and to the center post. The Φ obtained from these calculations was 2.12 x 10^{-5} Webers, compared with the 2.18 x 10^{-5} Webers yielded by our longhand permeance calculations, agreement to within less than two percent. In contrast, calculations of the flux density at Point G yield 0.39 T for the analogue method and 0.30 T for the computer, an approximately 30 percent agreement.

11

Table 2. Magnetic Flux Paths of Figure 4(a)

PATH DESCRIPTION	PATHS	FORMULAE
Gaps With Opposing Parallel Faces	P_{MG}, P_G, P_M	$P = xy/g$
Paths Between Parallel Linear Edges of Opposing Surfaces	$P^1_{GE}, P^2_{GE}, P^3_{GE}$	$P = 0.26y$
Annular Cylindrical Sections Extending Between Coplanar Surfaces	$P^1_{ML}, P^2_{ML}, P^1_{GL}, P^2_{GL}$	$P = 0.318\,y \ln(1 + 2w/y)$
Spherical Quadrants ("Orange Slices") Connecting Opposing Corners of Opposing Parallel Surfaces	P_{GC}	$P = 0.077\,g$
Semicylindrical Sections Backstreaming from the Sides of the Magnet	P^1_{MS}, P^2_{MS}	$P = 0.318\,y$
Quadrants of Spherical Shells ("Melon Slices")	P_{GB}, P_{MB}	$P = 0.25\,w$
"Orange-Slice" Sections Backstreaming from Longitudinal Edges of Magnet	P_{MC}	$P = y/16$

IN THE FORMULAE: x and y are dimensions of the rectangular gap faces perpendicular to the flux lines and g refers to gap lengths parallel to the flux lines respectively. w refers to thicknesses of cylindrical and spherical shells. The multiplying factors below indicate the number of flux paths of that particular shape.

Calculated Values of Permeances in Meters

$P_{MG} = 0.200$	$P^1_{GL} = 0.349$	$2 \times P_{GC} = 0.031$	$P^2_{MS} = 0.318$
$P_G = 1.000$	$2 \times P^2_{GL} = 0.140$	$2 \times P_{GB} = 0.100$	$2 \times P_{MC} = 0.050$
$P^1_{GE} = 0.260$	$2 \times P^1_{ML} = 0.052$	$2 \times P_{MB} = 0.050$	$P_M = 0.300$
$2 \times P^2_{GE} = 0.104$	$P^2_{ML} = 0.129$	$2 \times P^1_{MS} = 0.076$	$P^E_t = 2.901$
$2 \times P^3_{GE} = 0.042$			

$$P_t = 0.2719$$

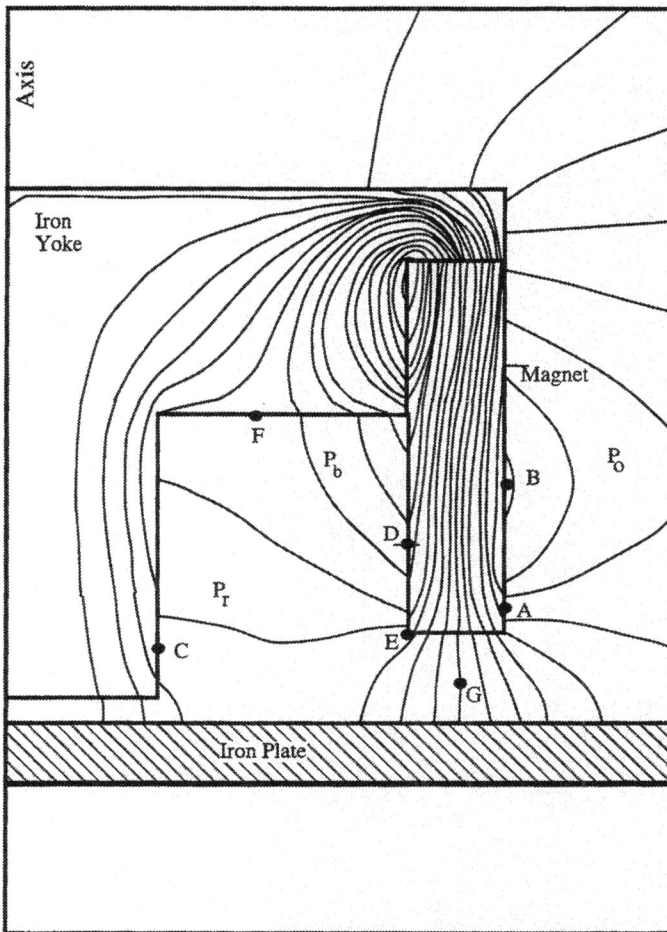

Figure 5. Computer plot of flux lines produced by a cylindrically symmetric speaker magnet of E-shaped cross section in proximity with an iron plate.

2.3 Flux Confinement by Cladding

Experience in magnetic design teaches us that, in most magnetic circuits, much of the flux generated by the permanent magnets is wasted by distribution among unwanted flux paths exterior to the work space of interest. The electric analogue method suggests a solution to this problem. In an electrical circuit, current can be kept from flowing along extraneous paths either by insulation or by

13

employment of compensating or bucking potentials, as in a potentiometer. Magnetic insulation is not practicable at present pending the arrival of room temperature superconductors with sufficiently high lower critical fields (H_{cI}). However, with rigid rare earth permanent magnets such as the NdFeB, the potentiometric method has proved to be very fruitful in magnetic design.

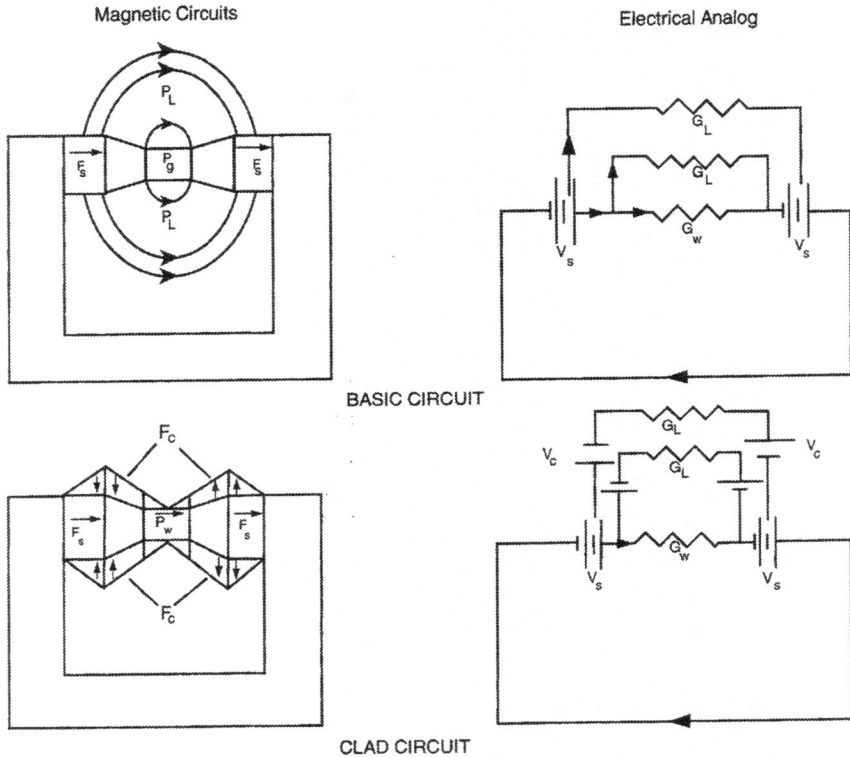

Figure 6. Gap field intensification through cladding.

In illustration, let us consider the simplest of magnetic circuits, the horseshoe-like configuration of Fig. 6. Clearly, much of the generated flux leaks into regions P_L. The equivalent current in the electric analogue would flow through wires G_L. The analogy is not exact, since the G_L are discreet, while the P_L are continuous. However, this slight discrepancy is irrelevant to illustrating the principle of cladding. In the electric circuit, batteries V_c of the appropriate EMF are placed in G_L, so that no current flows there. To effect the analogous suppression of flux in P_L of the corresponding magnetic circuit, it is clear that permanent magnets F_c analogous to the batteries V_c must be placed there. We begin by assuming that this goal has been accomplished and then work backwards to find the appropriate magnet configuration.

14

By hypothesis, the total flux furnished by the magnets flows exclusively within the cross section of the circuit and is given by the magnetic Ohm's law:

$$\Phi = F / R_t = -2L_m B_{0c} / (R_m + R_g) = 2 L_m B_r / (R_m + R_g), \tag{18}$$

where L_m is the length of a single magnet, B_r is the magnet remanence, R_m and R_g are the reluctances of the magnets and gap, respectively, and B_{0c} is the coercivity, which for our perfectly rigid magnets is equal to $-B_r$. The reluctivity of the magnets is assumed to be unity, and that of the yoke, zero. For total flux confinement to be effected, every point on the surface of, or exterior to the final configuration, must be at the same potential. This condition is fulfilled automatically for points on the equipotential surface of the iron yoke. Therefore, cladding must be placed around the magnets and the gap so that its outer surface potential is everywhere equal to that at the yoke surface. Therefore, the potential difference between any point C on the cladding surface and point A on the yoke must be equal to zero and given by the line integral along any path connecting A and C. We consider the path A B C shown in Fig. 7.

$$F_{ABC} = F_{AB} + F_{BC} = 0 \tag{19}$$

Now $F_{AB} = F_m x_B / 2L_m$, where F_m is twice the mmf across one of the magnets, and $F_{BC} = B_{od} y_{BC}$, where B_{od} is the radial magnetic field in the cladding and y_{BC} is the cladding thickness at point B. Since $F_m = \Phi R_m$ Eq. (19) becomes

$$0 = \Phi R_m x_B / 2L_m + B_{od} y_{BC} \tag{20}$$

Substitution of 18 in 20 yields

$$y_{BC} = F_{BC} / B_{od} = -F_{AB} / B_{od} = -\Phi R_m x_B / 2L_m B_{od} \tag{21}$$

Since, by hypothesis, no flux flows to the exterior, the radial flux density in the cladding magnet must also be zero. From the demagnetization curve of a magnetically rigid material, we know that at zero flux density, the field is equal to the coercivity $B_{0c} = -B_r$. Substitution of this value for B_{od} in Eq. (21) yields

$$y_{BC} = 2L_m B_r R_m x_B / 2L_m B_{0C} (R_m + R_g) = \frac{R_m x_B}{\left(R_m + R_g\right)}. \tag{22}$$

and the outer surface of the cladding about the supply magnet is a truncated cone, whose half angle is $\tan^{-1}[R_m /(R_m + R_g)]$. The cladding thickness y reaches maximum at the end of the magnet, which is at the gap edge. The thickness declines beyond the edge because the field in the gap B_{0g} is in the direction opposite to that of B_{0m} and it is given by

$$B_{0g} = B_g = B_m = \Phi / A_m , \tag{23}$$

15

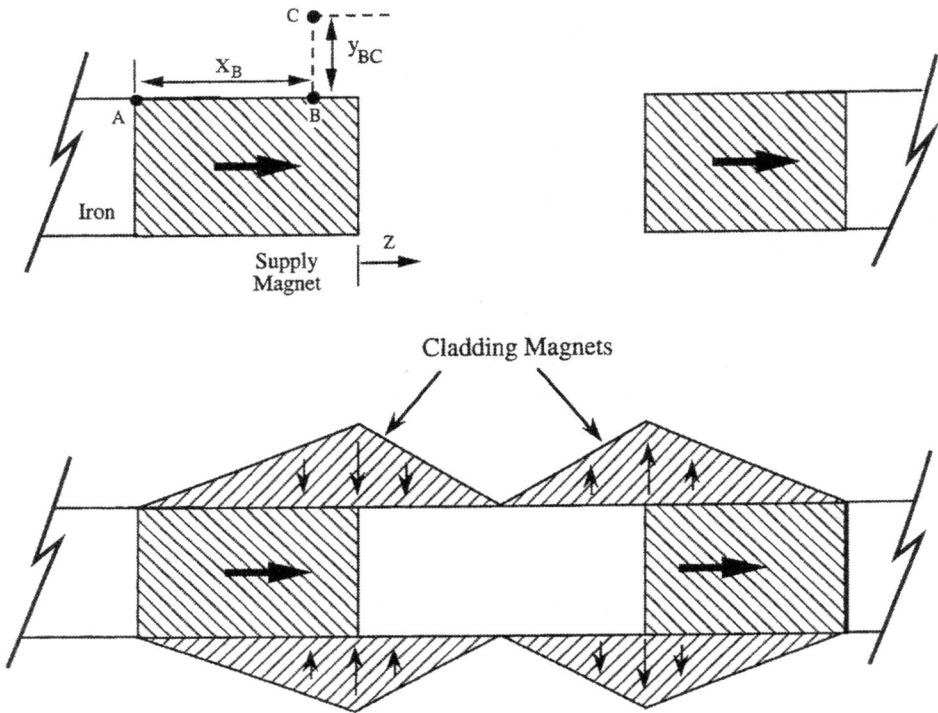

Figure 7. Determination of cladding thickness of a horseshoe magnet in the gap region.

where A_m is the cross-sectional area of the gap. The potential declines with distance, z, from the gap edge as

$$\Lambda F = -zB_{0g} = zB_m = z \, \Phi / A_m \tag{24}$$

Inserting (18) for Φ we obtain

$$\Lambda F = -2zL_m B_r / A_m \, [R_m + R_g] \tag{25}$$

At midgap, ΛF is equal to $-F$ at the gap edge so that F is zero there and takes on increasingly negative values beyond. The negative potentials are balanced by cladding of opposite polarity, which increases in thickness up to the other gap edge, after which it declines linearly to zero. The resulting antisymmetric structure is shown in Fig. 7. In practice, the gap cladding is sometimes omitted for easier

16

access. This results in a reduction in gap field and some bowing of the field lines which is often tolerable, but becomes more serious as the gap length, g, increases.

The efficacy of cladding is demonstrated by a consideration of the unclad structure and its identical clad counterpart in Fig. 6. The latter has a gap field of 2 T, as compared with only 0.8 T for the unclad structure. Mass efficiency of the cladding is demonstrated by the 7 kg mass of such an assembly compared to the 55 kg required to produce only 1.6 T in an unclad configuration. The 2.2 T delivered by the clad version also compares favorably with that usually delivered by electromagnets that are many times as large, and is generated without power supply or current source. Cladding is especially effective when the leakage permeances are a large fraction of the total external permeance. When the same horseshoe-like structure has no tapered pole pieces as in Fig. 7, most of the external permeance is in the gap, and cladding raises the gap field from 0.5 T to only 0.8 T.

2.4 Clad Permanent Magnet Solenoids

Traditionally, uniform magnetic fields extending over distances that are long, compared to the diameter of a cylindrical working space, are produced by electrical solenoids. These are cumbersome, requiring power supplies which often entail the expenditure of considerable energy. The rare-earth materials provide solenoidal fields with permanent magnet structures. A particularly useful and ingenious design for such a configuration was conceived by Neugebauer and Branch[7] to focus the electron beam in a microwave klystron (Fig. 8). The required flux density was to have a magnitude of 0.15 T in a direction parallel to the axis of a cylindrical space 21.6 cm long and 6.99 cm in diameter. The REPM magnet used to supply the flux is in the form of an axially magnetized annular shell with μ_r approximately equal to unity forming the perimeter of the work space. Abutting against its ends are disks of a passive ferromagnet of high permeability and saturation magnetization, such as iron or permalloy. These disks serve as pole pieces to guide the flux produced by the magnet into the cylindrical work space. Again, we require that all of the flux generated by the magnet pass through the work space. To accomplish this, the magnet must have a cross-sectional area A_m dictated by conservation of flux, viz:

$$B_m A_m = B_w A_w = B_{0w} A_w \tag{26}$$

where B_m is the flux density in the magnet and B_w that in the work cavity, and where A_w is the cross-sectional area of the cavity. If, as we assume, flux confinement has been accomplished, and since the end discs are equipotential surfaces, the field in the magnet B_{0m} must equal B_w. B_m can then be determined from (26) and from the equation for the demagnetization curve of a rigid magnet:

$$B_m = B_{0m} + B_R = B_w + B_r \tag{27}$$

Substitution of this expression into (26) yields A_m:

$$A_m = B_w A_w / (B_w + B_r) = A_w / (1 + B_r / B_w) \tag{28}$$

To find the cladding thickness necessary for flux confinement, we take our zero potential to be at the left pole piece in Fig. 8. For any point C on the outer surface of the cladding to be at zero potential, the line integrals of H from A to B and from B to C must be equal and opposite, that is

$$B_w x = -B_{oC} t_B \qquad (29)$$

Since no flux flows radially through the cladding, B_{mC} is zero; and the rigid-magnet demagnetization curve shows that B_{0C} is equal to the coercivity B_{0C} which in turn is equal to $-B_r$. Therefore, from (29) we see that the thickness, t, anywhere must be

$$t = B_w x / B_r \qquad (30)$$

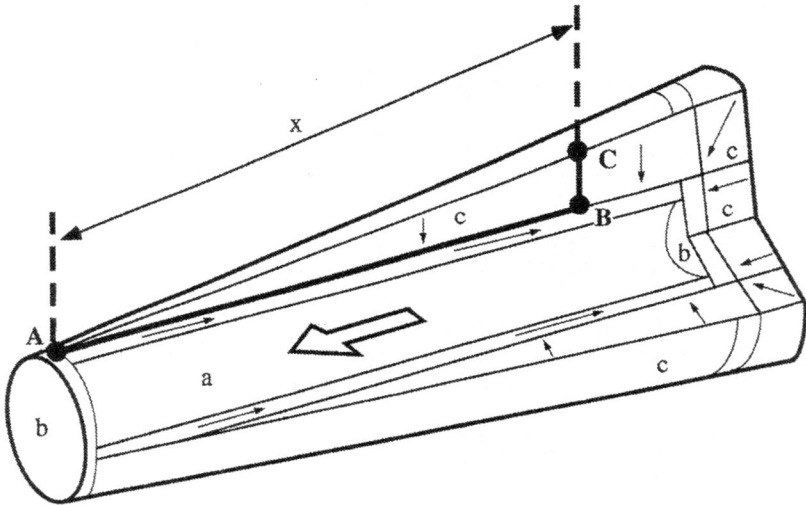

Figure 8. Permanent magnet solenoid of the Neugebauer-Branch type. a. Flux supply magnet. b. Pole pieces. c. Cladding magnets.

Again, t is linear in x and a conical surface results, as shown in Fig. 8. Because it is at the same potential as the right end of the supply magnet, the pole piece on the right must also be encased in cladding of maximum thickness.

The actual structure of Neugebauer and Branch employed two tandem chambers with oppositely directed fields, so that cladding of the bases of the high potential pole pieces was unnecessary; a single iron disc served as a common high potential pole piece for the two chambers.

A single-chambered structure can be made less bulky than the Neugebauer configuration by taking the plane midway between pole pieces to be at zero potential.[8,9] Then, the cladding is in the form of two conical structures of half the base thickness of the Neugebauer structure. (See Fig. 9.) The cladding to the left is oriented outward and that to the right inward. In the Neugebauer structure, shift of zero

18

potential from chamber end to chamber center results in an approximately 60 percent reduction in mass.

Figure 10 shows a computer plot of the magnetic flux produced by the improved structure. There is some leakage due to imperfect cladding at the ends. Figure 11 shows the axial field as a function of distance from a chamber end and the much improved field uniformity obtained after a simple correction to the cladding slope is made. The correction consists of an increase of cladding thickness Δt according to the formula:

$$\Delta t = (H_d - H_a)t / H_a \tag{31}$$

where H_d is the desired field and H_a is the actual field.

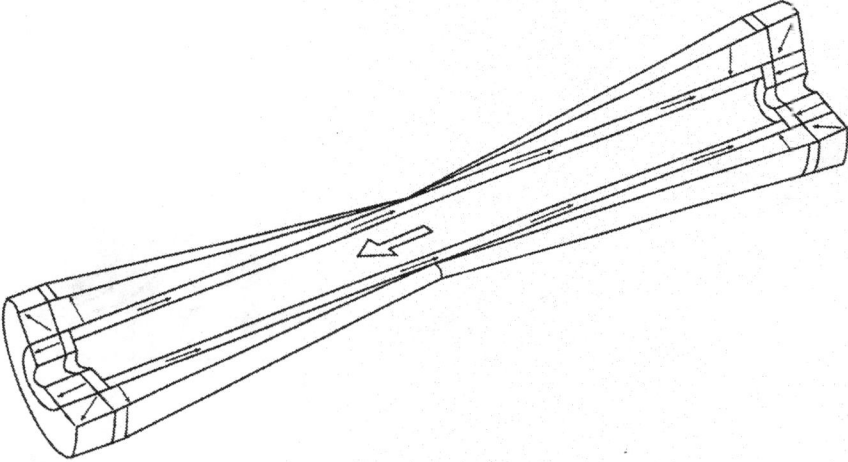

Figure 9. Solenoidal magnetic field structure with zero potential reference in center.

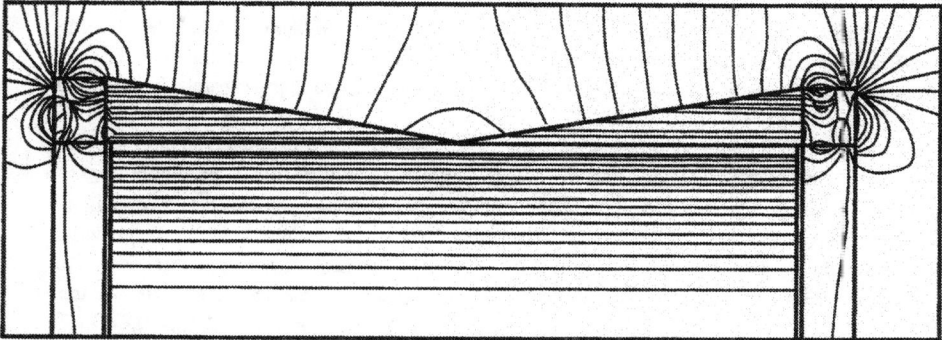

Axis of Rotation

Figure 10. Flux plot of magnetic field of the structure of Fig. 9. Apparent flux crowding towards the periphery is because each line represents a unit of flux in an annular ring of given thickness.

19

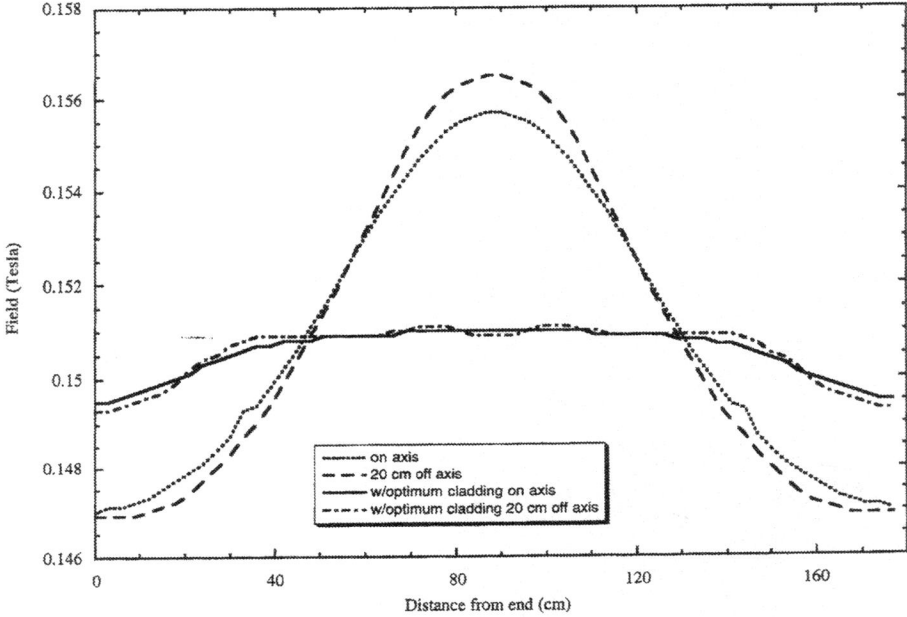

Figure 11. Axial magnetic field profile of the structure of figure 10.

Figure 12 shows the impracticality of alnico for permanent magnet solenoids. Since the cladding thickness is inversely proportional to the coercivity, it must be ten times as thick for alnico as for even the very modest properties of the REPM of Fig 12. This translates to an approximately hundredfold increase in mass.

There are many useful variants of the permanent magnet solenoid. For example, the internal field need not be constant but can be varied along the axis by appropriate adjustments in the supply magnet and cladding thicknesses.[8,10] As an example, if the field is to increase linearly with progression along the axis, the cross-sectional area of the supply magnet must also increase linearly while the cladding thickness is parabolic. In such a structure, the most efficient placement of zero potential is at the magnetic center C of the configuration, that is, where the line integral of the field along the axis over the length of the chamber attains half its value i.e., $\int_0^C \vec{H} \cdot d\vec{l} = \int_C^L \vec{H} \cdot d\vec{l}$ where 0 and L are the ends of the working space. Clearly C lies on the high field side of the geometric center and the cladding is concave on the high field side and convex on the low field side.

20

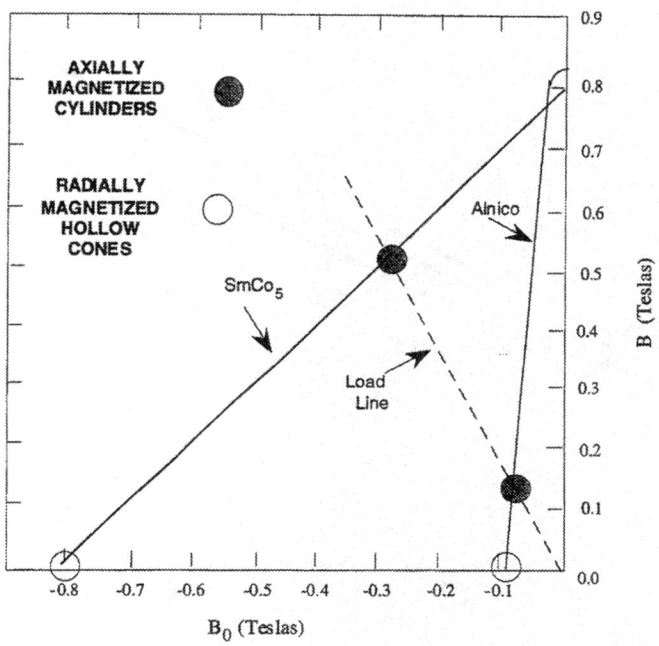

Fig. 12 Demagnetization curves and operating points for the magnets in a solenoidal field source.

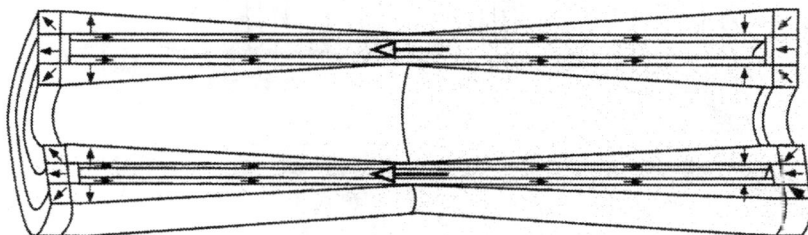

Figure 13 Permanent magnet annular field source.

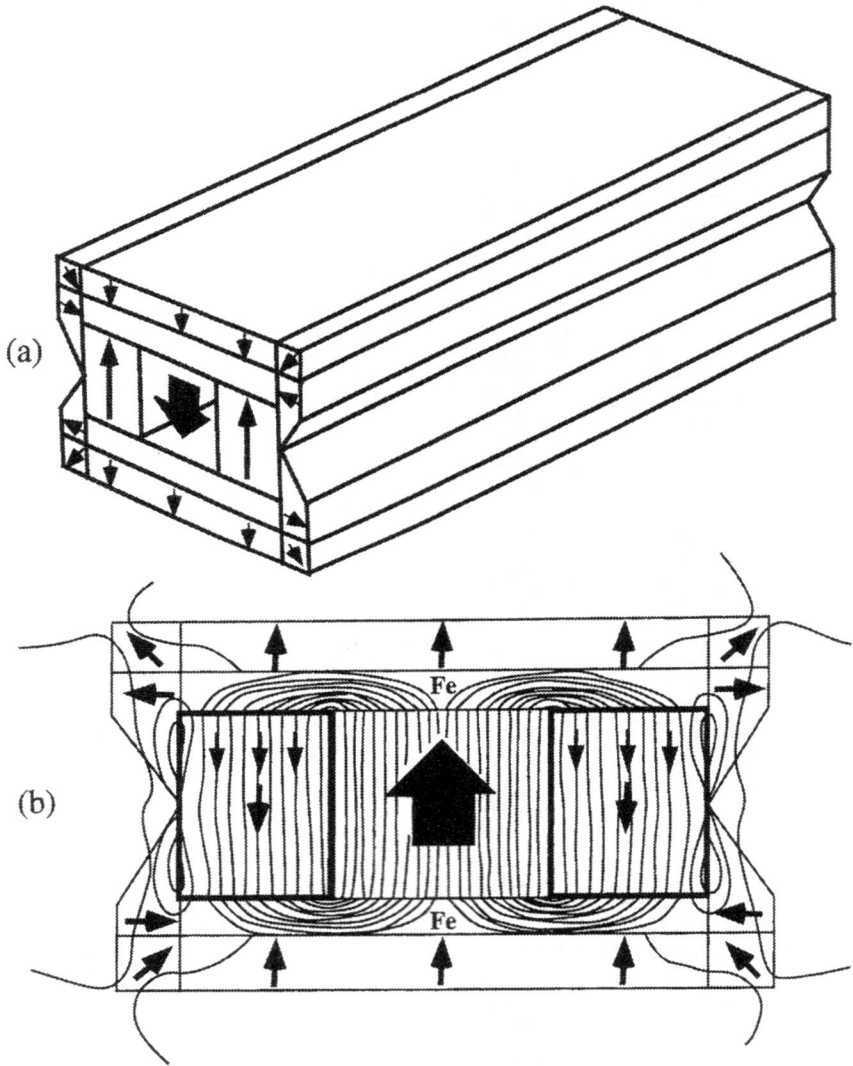

(a)

(b)

Figure 14. Transverse field source for a rectangular cavity(a) and its sectional flux plot(b) in the cavity. The large arrow indicates the field in the rectangular central cavity, the medium sized arrows the orientations of the permanent magnets and the small arrows, the directions of the flux lines. The supply magnets are marked by the heavy lines. All other magnets in the flux plots are cladding. Note the efficiency of the cladding in the flux confinement as evidenced by the negligible leakage only at the corners.

The structural variations necessary to obtain the desired fields can be effected parametrically as well as geometrically, i.e., material properties of the magnets can be varied rather than their dimensions.[8-10] This procedure is not as efficient with regard to weight and bulk of the structure because the uniform cladding thickness must be the maximum thickness in the geometric structure. On the other hand, the ring magnets are all of the same dimensions and are easier to manufacture.

The outlined principles can be applied to other field configurations such as in the cylindrical structure of Fig. 13 in which the field is confined to an annular ring by the placement of cladding on both the inner and outer boundaries of the working space.[11] In this arrangement, the supply magnets can be located at either the inner or outer boundary of the working space, or both. Of course the specified requirements must be such that the inner hollow is spacious enough to accommodate both the inner cladding and all of the supply magnet. Clad rectangular structures with fields normal to the axis can be produced as in Fig. 14.[9] Such configurations are useful for Magnetic Resonance Imaging (MRI). MRI application will be discussed more fully in the next section.

All clad structures lose some of their field uniformity if access holes are drilled into their interior cavities. Such holes can often be made to be axially symmetric so that the field distortions are also axially symmetric and can be at least partially compensated for by methods similar to that employed in the compensation illustrated in Fig. 11.

If only one hole is needed in a structure that has its zero of potential at an end, it is usually better to drill the hole at that end since then no penetration of a cladding magnet is necessary and there is no field reversal in transit.

3. Linear Structures Composed Entirely of Rigid Permanent Magnets

Iron flux guides or yokes to shape and intensify the field in a working space can be used with the alnicos but for many working-space geometries, it would be better to arrange magnetic poles so that maximum field is produced where it is wanted. Of course, there are no free magnetic poles; they are always accompanied by a like number of opposite poles of equal strength. It is the goal of yokeless magnet design to place the component magnets so that their poles are as effectively placed as possible to produce the desired field where needed while the opposite poles are more remotely placed or negated. Because the optimal arrays often entail very strong demagnetizing fields, the art of yokeless magnet design has very limited applicability with the older magnets of low coercive force.

Simple all-magnet structures that produce extraordinarily high fields can be obtained through considerations of symmetry and a very useful theorem.[12,13] The latter states that if the magnetization of an infinite line source oriented perpendicular to its axis is rotated about that axis, the field it produces remains everywhere constant in magnitude and is everywhere rotated by the same angle in the opposite sense. This is easily demonstrated. For an infinite line dipole

$$H_\theta = P\sin\theta / r^2 \tag{32}$$

$$H_r = P\cos\theta / r^2 \tag{33}$$

23

Therefore the magnitude $|H|$ is

$$|H| = \left(H_\theta^2 + H_r^2\right)^{\frac{1}{2}} = P/r^2 \qquad (34)$$

and is independent of dipolar orientation. P is the moment per unit line length and θ is measured from the dipolar axis. Equations 32 and 33 indicate that the field orientation angle α is twice θ. If we then widen θ by a rotation of the magnetization, α must change by the same amount in the opposite direction to preserve the relationship $\alpha = 2\theta$. This principle has been used extensively by Halbach in his design of various particle beam devices.[12-15]

As an example of this principle's application, we consider a cylindrical cavity of infinite length in which it is desired to generate a uniform field transverse to the principal axis. If this is to be accomplished with permanent magnet material in the form of a circumscribed cylindrical shell, it is clear that the orientation of magnetization must have reflection symmetry, as in Fig. 15. Also by symmetry, the magnetization of the infinitesimal sector, $\mathbf{A}$, must be radially oriented in the direction of the desired field.

If all the other sectors of the annular magnet ring were also pointed radially inward, each would contribute a field, dB_w, at the center of the cavity equal to that of $\mathbf{A}$ in magnitude, but oriented at an angle of θ to that of $\mathbf{A}$, where θ is the azimuthal angle. Of course, this would result in mutual cancellation of all the dB_w's. According to our theorem, if the orientation of each segment is now changed by $-\theta$, the fields of the individual segments would all be in the same direction everywhere and hence would add without cancellation. Therefore, the prescription of the theorem is that the orientation of magnetization γ be in the direction

$$\gamma = 2\theta \qquad (35)$$

The magnetic field in the cavity can then be found either by determination of the surface, σ_s, and volume, σ_v; pole densities arising from the magnetization $\vec{M}$ and insertion of the results into Coulomb's law; or by application of Maxwell's equations to the boundaries of the magnetic material.[17] The pole densities are given by $\sigma_s = \hat{n} \cdot \vec{M}$ and $\sigma_v = -\nabla \cdot \vec{M}$, respectively, where $\hat{n}$ is the unit vector normal to the surface under consideration. The resulting field in the cavity (w) is uniform and given by

$$B_w = B_r \ln (R_2/R_1) \qquad (36)$$

where R_1 and R_2 are the inner and outer radii and B_r is the remanence of the magnet. Thus it is seen that there is no limit to the magnetic field that can be obtained in such a structure provided only that the outer radius R_2 be made large enough. Due to the logarithmic dependence of B_0 on R_2, material bulk quickly becomes prohibitively large with relatively small field increments. Nevertheless, fields of twice the material remanence should be practicable, namely 2.0 to 2.5 T for the highest energy product materials used, e.g., Nd-Fe-B. If, for example, $B_r = 1.2$ T, the working space is 2.5 cm in diameter and the outer structural diameter is 15 cm, the internal field will be given by

24

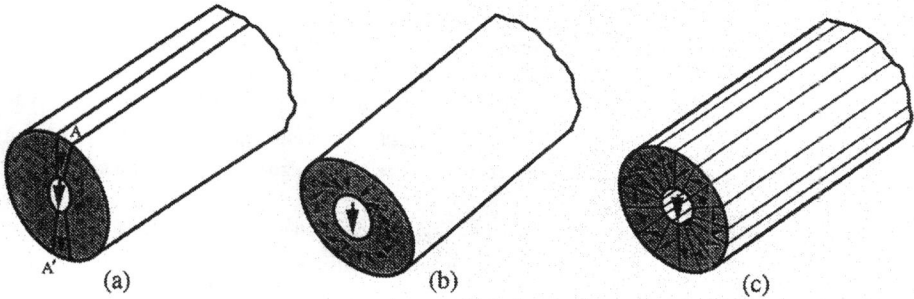

Figure 15. Cylindrical dipolar field source. (a) If a cylindrical dipolar source is to produce a uniform field in a cylindrical cavity as shown above, it is clear from symmetry that the infinitesimal elements A and A´ must be oriented as shown. The large arrow indicates the field in the inner cavity. (b) Ideal structure resulting from rotation theorem. (c) Segmented structure approximating (b).

$$B_w = 1.2 \ln (15/2.5) = 2.1 \text{ T} \tag{37}$$

which is an impressive field to be generated in so large a working space by a structure that can be packed into a cylinder 15 cm in diameter and which generates no stray field.

Ease of manufacture can be facilitated by approximation of the cylindrical boundaries with circumscribed polygons so that the structure is made of trapezoidal segments as in Fig. 15c. [13] If this is done, the expression for the field becomes

$$B_w = B_r \left[\frac{\sin(2\pi/N)}{2\pi/N} \right] \ln(R_2 / R_1) \tag{38}$$

where N is the number of sides of the polygons. Even polygons with as few as eight sides still generate a field 90% of that obtained with ideal circular cylindrical surfaces.

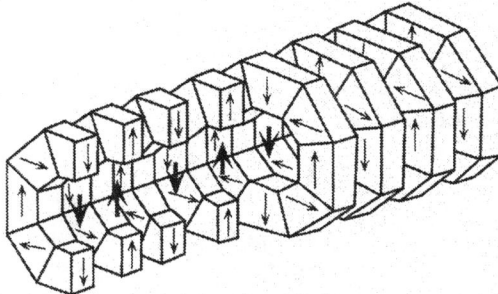

Figure 16. Wiggler made from slices of octagonal cylindrical transverse field sources.

25

Structures of this type are useful for any application in which high transverse fields are required in elongated cavities as in nuclear magnetic resonance imagers, biasing fields in electronic filters and general utility laboratory magnets. Cross-sectional slices of the structures can be used as segments of wigglers and undulators of free electron lasers to provide higher fields with less bulk and mass than is possible with conventional structures (Fig. 16).[12-16] Use of such structures and of their quadrupolar forms in particle beam guidance and MRI has been pioneered by Halbach[12-15], Holzinger[16] and Zijlstra.[17]

The cylindrical structures can also be made to be mechanically adjustable.[19] This is done by fabrication of the magnet in two rings with one nested inside the other as in Fig. 17. Then the field within the central cavity will be the vector sum of the fields produced by the individual rings. If the rings are dimensioned so that each produces one half the field individually and are rotated with respect to each other by an angle α, the magnitude of the field within the cavity is just

$$B_w(\alpha) = B_w(0) \cos(\alpha / 2) \qquad (39)$$

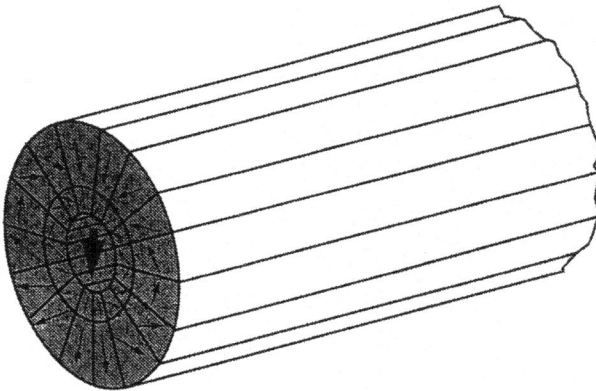

Figure 17   Adjustable circular-cylindrical transverse field source.

where $B_w(\alpha)$ is the maximum field, which occurs at $\alpha = 0$. Thus any field in the range $\pm B_w$ is easily available. For field equality of the individual rings, the radius R_3 of the boundary between them must be the geometric mean of the outer and inner radius of the device:

$$R_3 = \sqrt{R_1 R_2} \qquad (40)$$

Variable flux sources of the type shown in Fig. 17 can be approximated with only eight segments in each ring. They are very useful in laboratory instrumentation where a variable field is required. For example, a compact vibrating sample magnetometer has been described where a variable field of 0 to 1.2 T is generated in a 26 mm bore by a Nd-Fe-B structure.[20, 21]

Fields of higher multipole value than dipole can also be generated by an appropriately faster change

26

in variation of orientation, γ, with ϕ, so that

$$\gamma = ((n/2)+1)\theta \qquad (41)$$

where n is 2 for a dipole, 4 for a quadrupole, and so on.[13]

To make a ring of this type it is not necessary to make the many pieces with different orientations individually.[22] For example, in the manufacture of a dipolar source, it is only necessary to orient a ring in a single unidirectional field, as in Fig. 18a. The ring is then cut into as many sectors as the desired approximation to a perfect ring demands. The pieces are interchanged according to the prescription $\theta \leftrightarrows -\theta$ as shown in Fig. 18b and glued together to form the desired structure in Fig. 18c. This is possible because every possible orientation of magnetization with respect to the local radius is contained in the structure of Fig. 18a. They need only be reassembled to yield the dipolar source of Fig.18c. The same result can also be obtained by a 180° rotation of each segment about its local radius.

A similar procedure can be used to make higher pole sources, but then n structures such as that of Fig. 18a are required to yield the n of the desired n-poled arrays. The sequence of assembly of a quadrupolar source is shown in Fig. 19.

4. Flux Confinement to Polygonal Cavities

It is often desirable to produce a strong transverse magnetic field in a cylindrical cavity of equilateral polygonal cross section. It is sometimes further required that the magnetic field be completely confined to within the outer boundaries of the structure with said boundaries being parallel and similar to the cavity boundaries. The latter requirement is to ensure the possibility of successive nesting of similar structures within each other to take advantage of the additive property of such fields to produce a large interior field equal to the sum of the components. Because of its usefulness in MRI,[19,23] a square cross section due to Abele will be considered as an example of the method of determination of the structure.

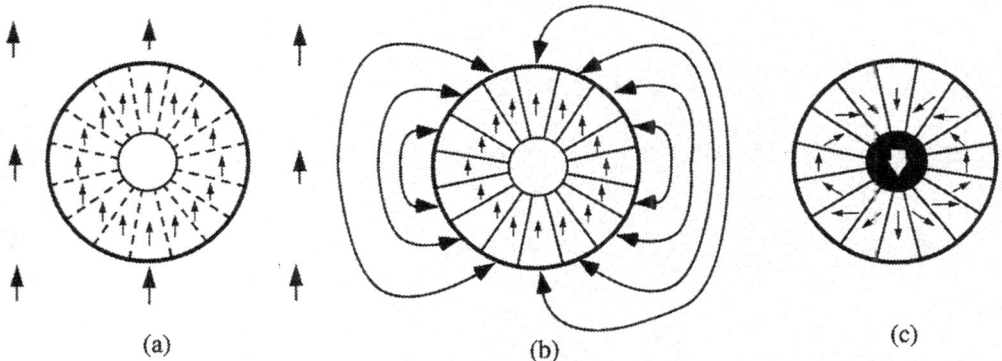

(a) (b) (c)

Figure 18. Construction of circular-cylindrical transverse field source. (a). Structure is magnetically aligned (small arrows) in a uniform magnetic field (big arrows) and sliced into segments (dotted lines). (b). Pieces are interchanged as shown and bonded to form the finished structure (c).

27

We begin with a consideration of the top segment of the square magnet. It is of trapezoidal cross section with 45° base angles. Because of square symmetry, only one half of it need be considered. For simplicity, it is desirable to have fields, magnetization and flux density constant over any of the segments into which the half trapezoid is to be divided. It is not possible to accomplish this together with flux confinement and generation of the desired field with a single piece. Accordingly, to attempt to fulfill all the requirements, we divide the segment into two triangular pieces 1 and 2 as shown in Fig. 20. The strength of magnetization $J = B_r$ is the same for both. Since it is clear that the orientation of the magnetization in segment 1 should be in the direction of the desired flux density B_w, we need only determine the thickness t and the orientation of the magnetization in segment 2 that will fulfill the requirements:

(1) That there be no flux exterior to the structure.

(2) That the specified uniform flux density, B_W, be generated in the interior cavity.

(3) That the exterior boundary of the structure be a square.

The first condition applied to the circuital form of Ampere's law dictates that

$$\oint H.dl = 0 = \mu_0 \oint H.dl = \oint B_0 dl$$

$$B_{01}y + B_{02y}(t-y) + B_w\,r_1 = 0 \tag{42}$$

or if paths of integration **ab** or **cd** of Fig. 20 are chosen

$$B_{01}t = B_{02y}t = -B_w r_1 \tag{43}$$

so that

$$B_{01} = B_{02y} \tag{44}$$

The second condition applied to Gauss' law at the cavity boundary yields

$$B_1 = B_w \tag{45}$$

And since

$$B_1 = B_{01} + J = B_{01} + B_r \tag{46}$$

Eq. (45) becomes

$$B_{01} = B_w - B_r \tag{47}$$

and insertion of (47) into (43) determines t viz

28

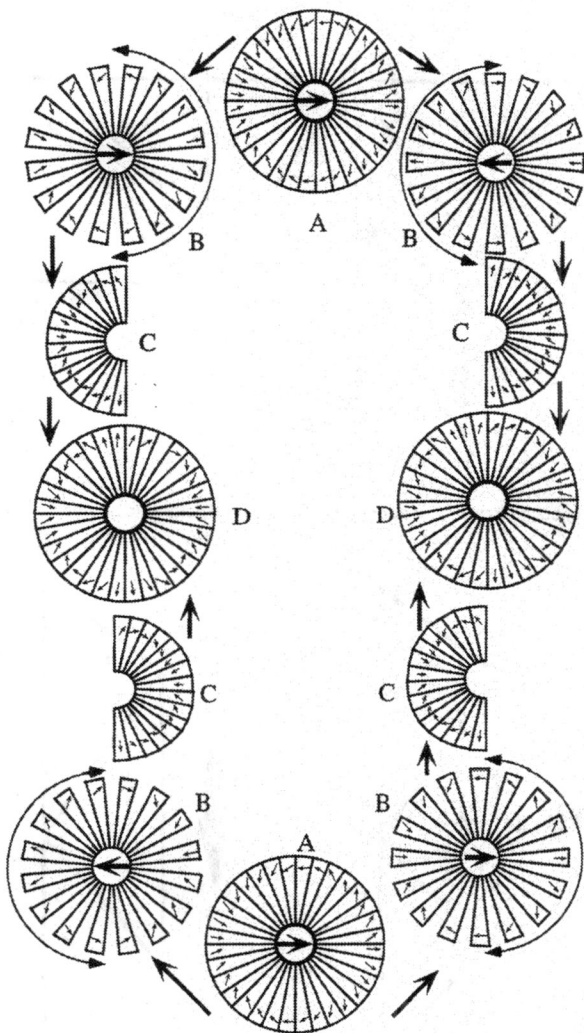

Figure 19. Formation of two quadrupole sources from two dipole sources. A. Start with two dipolar structures A. and B. Separate every other segment from adjacent segments to form two new structures for each original structure. C. Compress structures B in direction of circular arrows to form structures C. D. Assemble structures C to form structures D.

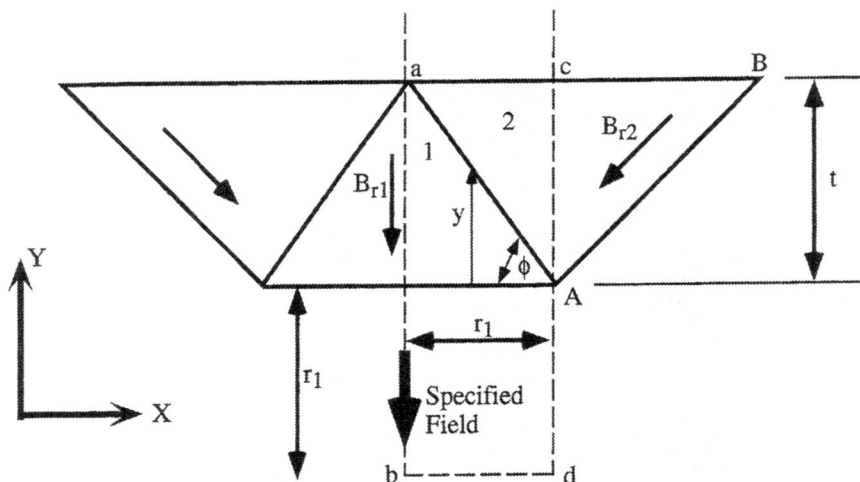

Figure 20. Determination of configuration of upper segment A square dipolar field source.

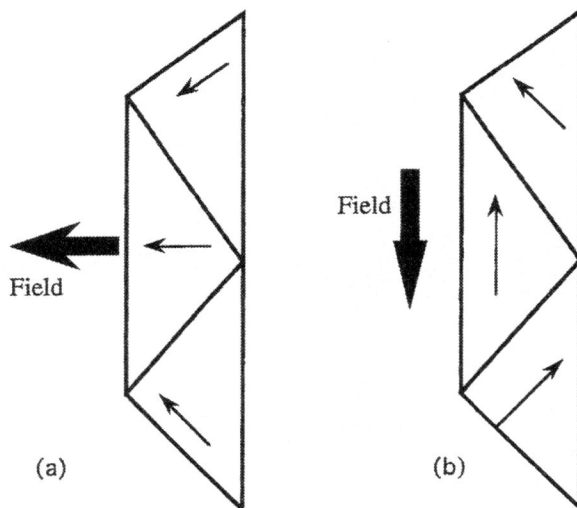

Figure 21. (a) Side segment of square if relative magnetic orientation were that of Fig. 20. (b) After -90° rotation of all magnetizations.

$$t = \frac{-B_w r_1}{B_w - B_r} = -\frac{r}{B_w/B_r - 1} r_1 \qquad (48)$$

$$t = r r_1 /(1-r) \qquad (49)$$

where $r = B_w / B_r$, that is, the fraction of the remanence flux density of the magnet that is manifested as useful flux density in the cavity. The fraction r cannot be chosen arbitrarily if the outer boundary is to be a square. To demonstrate this, we apply Ampere's and Gauss' laws to the outer surface. For **B** and **H** to be zero everywhere exterior to the outer boundary, these laws require that $B_{02X} = 0$ and $B_{2Y} = 0$ so that B_{02} and B_2 must be mutually perpendicular and respectively normal and parallel to said boundary and since $\vec{B}_{02} + \vec{B}_r = \vec{B}_2$

$$(B_{02})^2 + (B_2)^2 = B_r^2 \qquad (50)$$

Since the outer boundary is a square with its sides parallel to those of the inner boundary, application of Gauss' law at the boundary (1,2) yields

$$B_2 \sin \phi = B_1 \cos \phi \qquad (51)$$
$$B_2 = B_1 \, ctn \, \phi \qquad (52)$$

where ϕ is the angle between (1,2) and the inner boundary.

If we insert (51) and (52) into (50), we get

$$B_1^2 ctn^2 \phi + B_{02}^2 = B_{r2}^2 \qquad (53)$$

but

$$B_{01} = B_{02y} = B_{02} \,, ctn \, \phi = r_1 / t, B_1 = B_w \text{ and } B_0^1 = B_w r_1 / t \qquad (54)$$

so that (53) becomes

$$2 B_w^2 (r_1 / t)^2 = B_{r2}^2 \qquad (55)$$

$$2 (r r_1 / t)^2 = 1 \qquad (56)$$

$$\sqrt{2} \, r r_1 = t \qquad (57)$$

From (49) and (56) we get

$$r = \frac{\sqrt{2} - 1}{\sqrt{2}} = 0.293 \qquad (58)$$

31

If (58) is inserted into (56) we obtain the ratio of r_1 to t,

$$\frac{r_1}{t} = \frac{0.707}{0.293} = 2.41 \tag{59}$$

Because from (50), (54) and (55) we have that $B_{02} = B_2$, and the angle α between $\vec{B}_{r2}$ and $\vec{B}_2$ is

$$\alpha = \tan^{-1} \frac{B_{02}}{B_2} = 45°$$

so that the direction of $\vec{B}_{r2}$ is parallel to the boundary **AB** of Fig. 20. By symmetry, the directions of magnetization in the side forming the bottom of the configuration are just the inverse mirror images of those just formed for the top side. The magnetization orientations in the other two sides are found by the dipolar rotation theorem. If the side shown in Fig. 21 had its magnetization vectors arranged relative to its sides as in the top side, the resulting field would point to the left as in Fig. 21(a). For the field to re-enforce that of the top side it must be rotated 90° to make it point straight down as in Fig. 21(b) and therefore the magnetization must all be rotated -90°. The complete resulting configuration is shown in Fig. 22. A permanent magnet material of remanence 1.0 T produces a flux density of $B_w = 0.293$ T in the square cavity and there is no limit to the intensity that can be attained by successive circumscription of similar structures so that the total flux density is 0.293n T where n is the number of layers used.

Such square structures are particularly useful as MRI magnets because they can be compensated for small random defects incurred in manufacture and assembly.[24] This is accomplished by the placement of small dipoles at the inner or outer periphery of the structure, usually in the inner corners. The magnet is constructed from sectional slices that are compensated individually. The field that an ideal slice should generate is calculated and then compared with the measured actual field. The deviation field is then Fourier analysed to determine the magnitudes and directions of the compensating dipoles needed. Fortunately, satisfactory compensation can usually be achieved with terms no higher than the dipolar. See references 24, 25 for details.

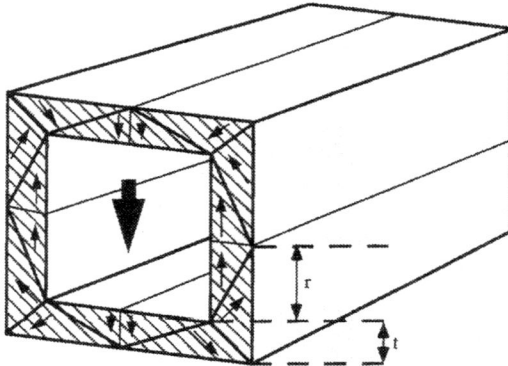

Figure 22. Square dipolar field source. Large arrow shows field in cavity. Small arrows show orientations of constituent magnets.

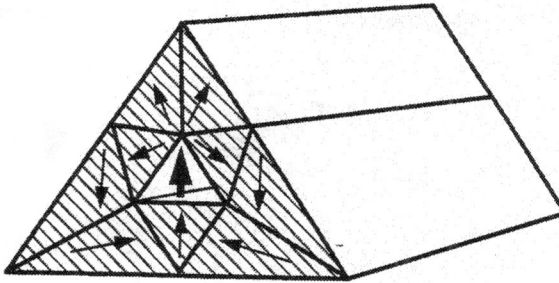

Figure 23. Triangular dipolar field source. Large arrow shows field in cavity. Small arrows show orientations of constituent magnets.

Table 3. Comparison of cylindrical confined flux sources with similar inner and outer boundaries. $B_w(T)$ is the field produced in a 1 cm cavity, $A_T(cm^2)$ is the cross-sectional area of the structure, $A_M(cm^2)$ is the cross-sectional area occupied by material, and B_r is the 1 T magnetic remanence used in all cases.

Structure	B_w	B_w/B_r	A_t	A_m	B_w/A_t	B_w/A_m
Triangular	0.50	0.50	5.2	3.9	0.96	1.3
Square	0.29	0.29	2.0	1.0	1.5	2.9
Octagonal	0.32	0.32	1.7	0.83	1.9	3.9
Circular	0.35	0.35	1.6	0.78	2.2	4.5

The same procedure can be used to find a triangular structure such as that of Abele[26] which fills the same requirements as those specified for the square structure (Fig. 23). Table 3 compares the triangular, square, octagonal, and circular configurations with regard to structural parameters. The triangular structure produces a higher field with a single layer than does the square. The circular structure produces a higher field than either with the same amount of magnetic material but requires more pieces for a near approximation (99%) to its theoretical performance. Therefore, if simplicity of manufacture is the prime consideration, the triangular section is best, but it is the worst in field strength produced for a given quantity of magnetic material. The circle is best in the latter regard but, as noted, is the most complex. The square is intermediate and can be easily dipole-compensated for structural defects. The octagonal approximation to the circle is simple and produces a strong field but does not afford total flux confinement.

Uniform fields can be provided in cylindrical cavities of any cross section. The shape need not be symmetrical or even externally convex. The details of the generation of such fields can be found in references 23-29.

If the required fields are less than about half the remanence, open compact sources for fields in axially finite cavities can be made from such sections and the open faces then clad according to the same principles described in the discussion of permanent magnet solenoids. A clad section of an octagonal source is illustrated in Fig. 24.

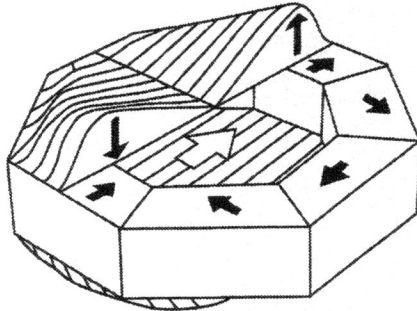

Figure 24. Section of an octagonal pyx. The source of the field is the octagonal ring. The rest of the magnets are cladding to keep the field in the interior (large arrow) confined and uniform.

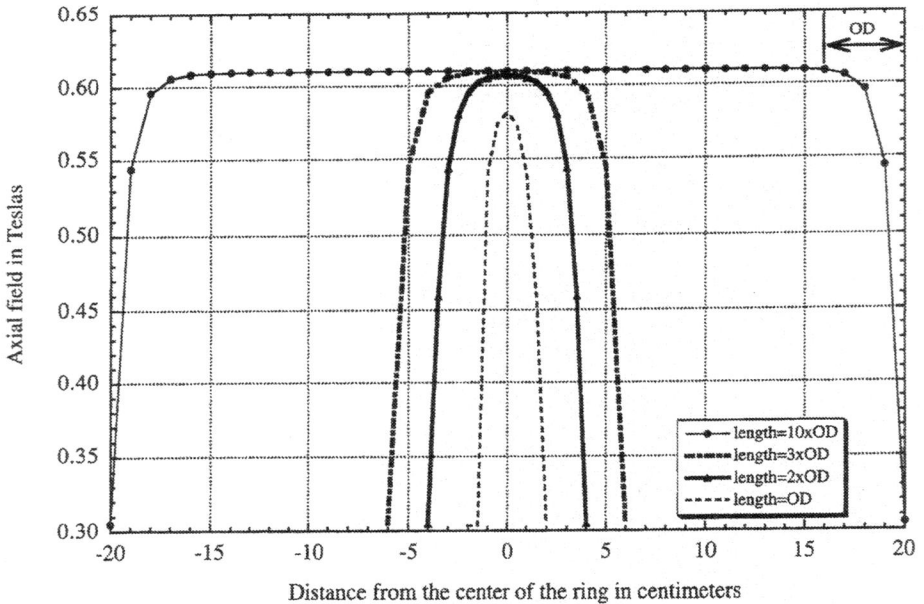

Figure 25. Transverse field along the axis of an octagonal approximation to a magic ring, with I.D. = 2 cm, O.D. = 4 cm.

End effects arise in unclad structures of finite length.[27] This is illustrated in Fig. 25 for octagonal magic cylinder approximations of different length, l. The field is essentially that of an infinite cylinder except where an end is approached to within a distance approximately equal to the structural diameter, d, where they begin to decline reaching about one half maximum at the end. For $l=3d$, the constant portion of the curve is still essentially at full field. For $l=2d$ there is no region of constant field but the maximum is nearly as great as in the longer structure. For $l=d$ the maximum field is smaller. Similar end effects are present in transverse field structures of different cross-sectional geometry.

Since permanent magnet MRI imagers are open ended, it would seem that their axial length would have to be prohibitive to provide a sufficiently large region of space with the required magnetic field uniformity. Abele[27] has provided an ingenious solution wherein by a judicious removal from the magnet cross section of slices of the proper thickness and location, the field harmonics that constitute the end effects are largely removed. A typical case cited is for a magnet with a length of twice its bore in which the field inhomogeneity over a central sphere of one quarter bore diameter is reduced by two orders of magnitude by the removal of five slices.

Other examples and details of the procedure can be found in reference 27 beginning on page 336.

In finite structures, the rotation theory holds only approximately, because of the change in problem dimensionality. As a result, in structures of squat aspect ratio, the formula $\gamma = 2\theta$ does not yield the most efficient configuration,[17] but the approximation is still a good one for $l/d > 3$.

The fields in all such arrays can be made to taper in magnitude with progression along the axis. The field plot in Fig. 26 is for one in which the field has a linear taper which is imposed by a like taper in the magnetic remanence of the tube. The graph shows end effects similar to those of the taper free structures. Tapered fields of this type are useful in MSW channelizers.

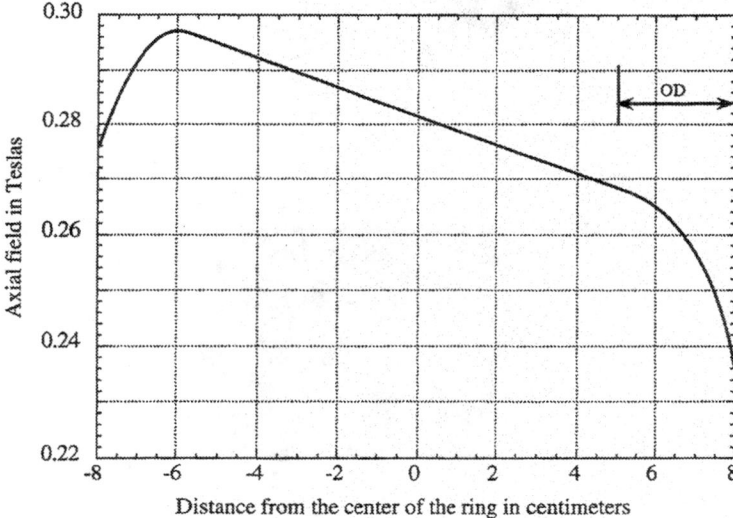

Figure 26. Transverse field along the axis of an octagonal approximation to a magic ring with a tapered field.

5. Rod Ensembles

Particularly useful but simple configurations can be assembled from transversely magnetized rods arranged in circular, cylindrical arrays.[20, 21] Figures 27a and 27b show two such configurations of four and six rods respectively. The field magnitudes can be varied by rotation of the individual rods and the field profiles of Fig. 28 are for the maximum and minimum fields obtainable for the four rod configuration or the magic "mangle." The mangle has been used by Coey et al. to apply magnetic fields to specimens examined on optical microscopes. The two prototype models produce variable fields over ±0.035T and ± 0.25T and uniform over a volume of radius 3 mm.

The more rods that are in such an ensemble, the more closely the field approximates that of a magic cylinder. For example, the qualitative resemblance of the axial field of the six rod "magic colt" of figure 29 to that of a magic cylinder is obvious.

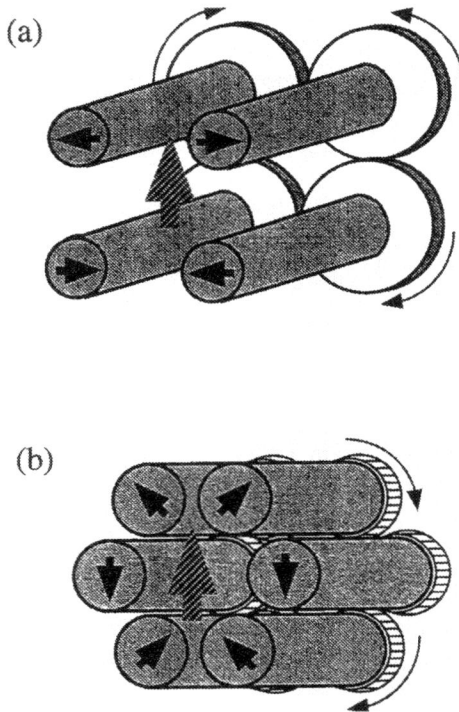

Figure 27. Variable flux sources composed of four and six rotatable rods. The orientations shown correspond to the maximum field position.

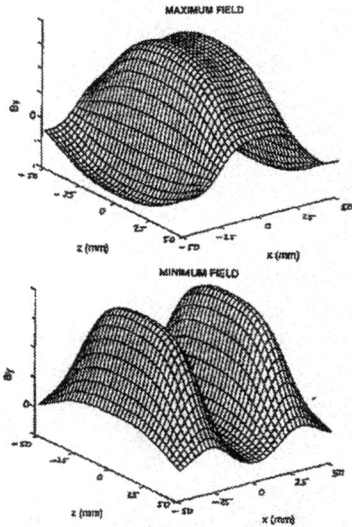

Figure 28. Field profiles for the structures of Fig. 27.

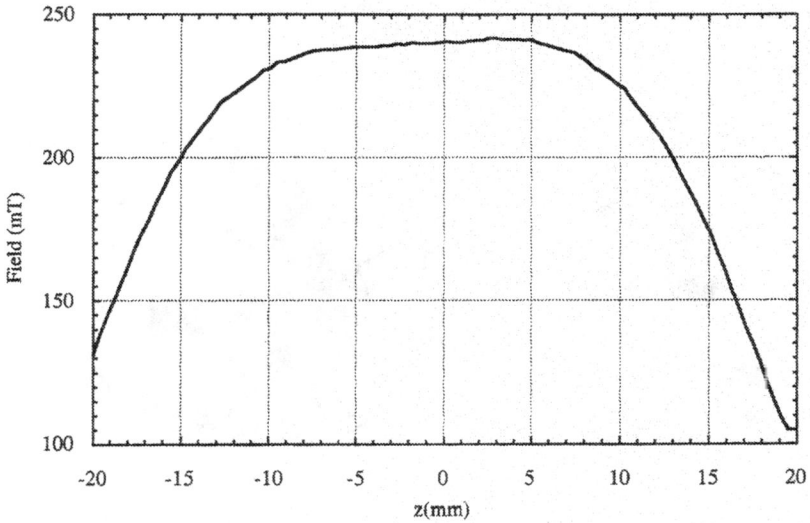

Figure 29. Field profile for the 6-rod magnet or magic colt configuration at maximum field.

6. Three Dimensional Structures

If a thin section of any cylindrical transverse flux source is rotated 180° about its polar axis, its locus forms a closed chamber in which, of course, there are no end effects as in a finite cylinder and in which the fields are greater than in the corresponding cylinders. For example, the spherical source pictured in Fig. 30a produces a uniform field in its cavity that is four-thirds as great as that in a cylinder with the same inner and outer radius.[17,22] If B_r is 1.2 T and a flux density of 2.0 T is desired in a cavity 2.5 cm in diameter, the outer diameter of the sphere need be only 10 cm. Figure 31 shows the fields attainable as a function of the outer radius for different values of B_r. This figure also illustrates the importance of obtaining higher B_r's for the production of higher fields at still lower material masses than the very affordable ones already available.

The fields produced in the spherical cavities are not only high but remarkably uniform along the axis even when axial tunnels are bored through the poles. Figure 32 shows the computer calculated field profiles for a sphere with $r_2/r_1 = 2$ and with axial tunnels of varying diameter. It is clear that for tunnels up to one-fourth the internal cavity diameter, the internal field profile is negligibly affected in amplitude and only slightly in uniformity. This suggests great utility for short Faraday rotators and possibly in-tandem periodic structures for particle beam focus. The latter is attractive because the field in the tunnels can be made to be nearly the reverse of those in the cavities. Because the square wave field configurations have higher axial field amplitudes than are possible in conventional periodic magnet stacks and because the squareness results in an average field value close to maximum, better focussing should result than from the conventional sinusoidal configuration.

Ideal Structure

Actual Structure

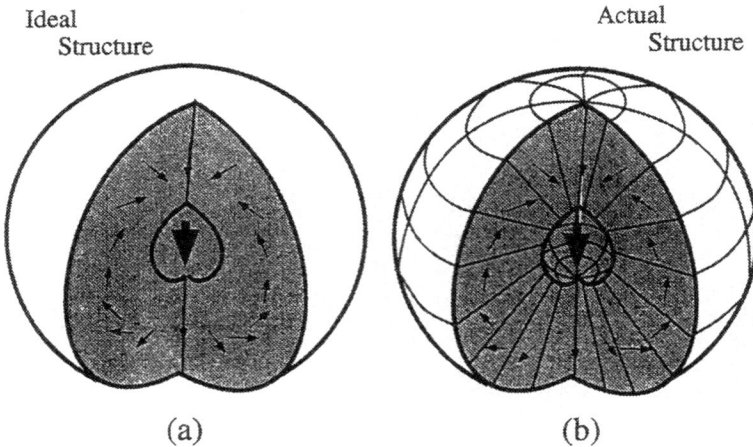

(a)   (b)

Figure 30. Spherical magnet structure. The field in the central cavity is given by $B_0 = (4\,B_r/3)\ln\,(r_2/r_1)$ where B_r is the magnetic remanence and r_2 and r_1 are the outer and inner radii respectively.

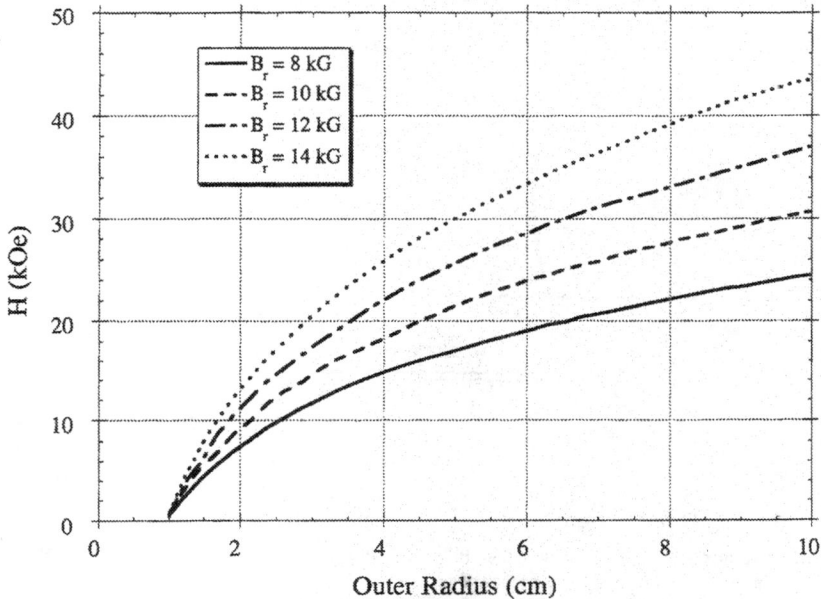

Figure 31. Plot of field vs outer radius for the magic sphere with 1 cm cavity.

A drawback of the spherical structure is that it does not confine magnetic flux completely,[19] but generates a small dipolar field outside of its outer surface. If this field is troublesome for a particular application, it can be easily cancelled by means of a uniformly magnetized shell on the outer surface of the spherical source. The shell is mounted with its magnetic orientation in opposition to the external field produced by the spherical source and is given just sufficient magnetic moment to cancel it. Since a uniformly magnetized spherical shell produces no field in its interior, it does not alter the specified field produced by the spherical source in the working space.

As in the permanent magnet solenoid, longitudinal field taper can be built into both magic spheres and magic cylinders.[29] Because of the very high fields and gradients produced by these structures they seem promising for use in ore and slurry separators where magnetic species are to be extracted. The figure of merit of such devices is the product of field strength with its gradient. The former induces large moments in particles of passive species and the latter provides the force that acts on those moments to pull the particles through the medium.

The gradients are provided parametrically with the remanence of the cylinder or sphere varying directly with the polar angle θ. Such a remanence provides a corresponding linear field gradient in the direction of the polar axis in the sphere and the polar plane in the cylinder.

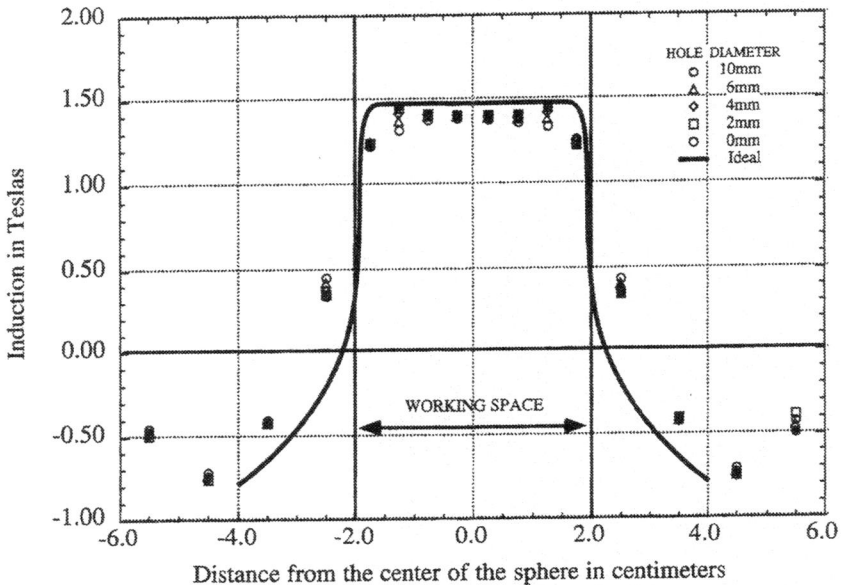

Figure 32. Axial field profile of the magnet structure of Fig. 30. Note the constancy of field in the cavity and also the relative insensitivity to the diameter of the access holes.

Magic spheres can also be made to produce even higher fields than those of Fig. 31 if only a portion of the interior cavity is needed for a working space. For example, the structure of Fig. 33a requires only a thin tunnel-like structure as in a Faraday rotator or magnetometer. In that case, the field in the tunnel can be augmented by a spherical permanent magnet in the unused portion of the cavity oriented counter to the field of the sphere. The resulting demagnetization field is then added to the original field. For a solid sphere, the augmenting field is $B_r/3$ which is over 0.4 T for some materials.

The structure of Fig. 33b is suitable for a working space in the form of an equatorial slot. In this case, the unused space in the cavity is filled with a passive ferromagnetic material such as iron which, if saturated by the magic sphere, adds to the original field a value of $2B_r/3$ which for some iron alloys is as great as 1.6 T.

Such structures are suitable for various particle beam devices such as synchrotron radiators; or free electron lasers - if the iron is inscribed with radial notches on the slot faces to give rise to an azimuthally periodic field.

If a working cavity of less extreme aspect ratio is needed, both of the above augmentation methods may be combined to form a cylindrical cavity of any desired aspect ratio. Figure 34 illustrates such a structure and Fig. 35 shows how the field at the center of the cavity varies as aspect ratio at a constant volume of 101 cm^3, an inner radius of 6.6 cm, an outer radius of 26.6 cm and a B_r of 1.2 T. The extreme fields attained are those of an equatorial slot and of an axial tunnel, 3.3 T and 2.5 T respectively. Figure 36 compares the various magic sphere types with regard to fields obtainable as functions of their outer radii. All types offer unprecedently high fields in proportion to structural size. For example, a design projected for use in a microwave tube employs a doubly augmented magic sphere with a cylindrical working space of aspect ratio 1.35. It produces a field in that space of 0.8 - 1.0 T depending on the B_r of the material chosen. Its mass is an order of magnitude less than those of electromagnetic or conventional permanent magnet designs and its iron pole pieces offer shielding of the field-sensitive electron gun as well as field distortion-free access for the rather large gun chamber and collector. Figure 37 shows the access of a standard sphere as compared to that of a doubly augmented

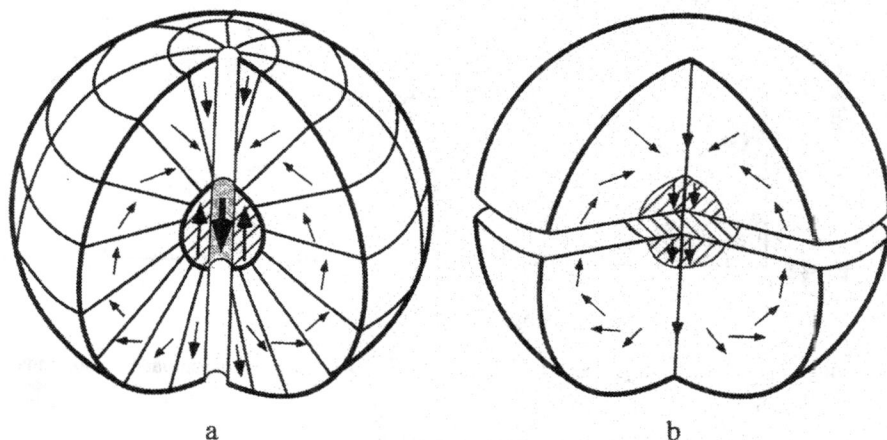

a   b

Figure 33. Two types of augmented magic sphere. (a) Tunnel working space, (b) equatorial working space.

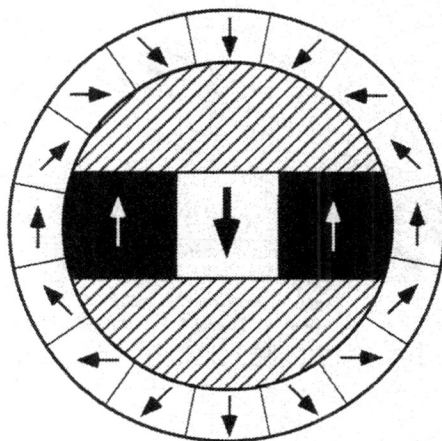

Figure 34. Doubly augmented magic sphere; ▨ Iron, ■ Permanent Magnet and ◺ Magic Sphere.

41

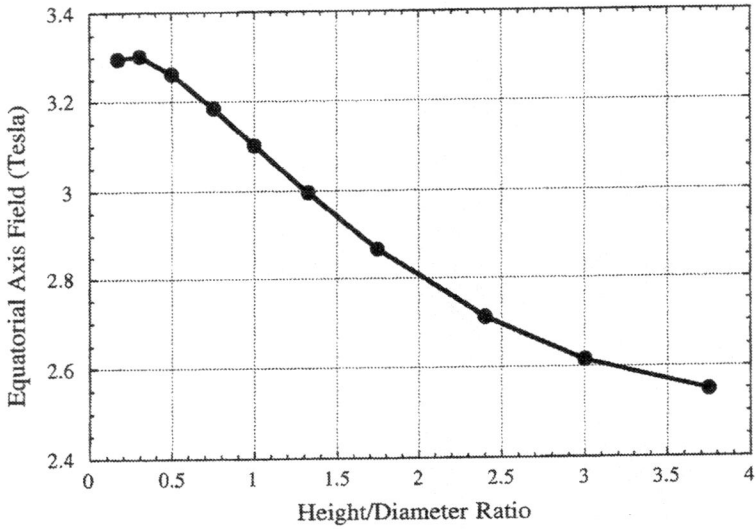

Figure 35. Range of field values in 101 cm^3 cylindrical working spaces of different aspect ratios in augmented magic sphere of r_o / r_i of 4.

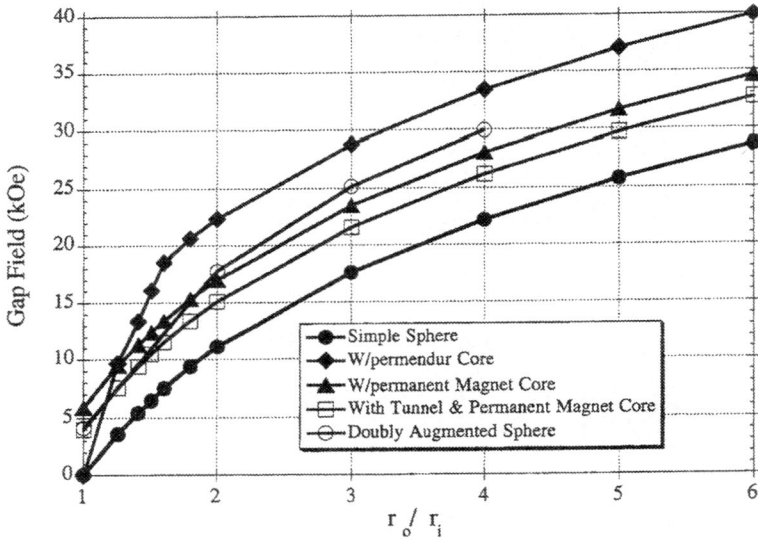

Figure 36. Fields of different types of magic spheres.

42

one. Figure 38 shows the contrast in field distortion caused by polar penetration of the two designs. The doubly augmented sphere shows near-zero field in the gun while producing a constant field of the required magnitude in the working space which is large enough to contain the microwave structure.

Unaugmented magic spheres can also be altered to provide relatively distortion-free access via the following consideration. A uniformly magnetized spherical shell produces no magnetic field in its interior. If, therefore, the magnetization of such a spherical shell is vectorially added to that of a magic sphere, the resultant would produce the same field in the interior as the magnetization of an unaltered

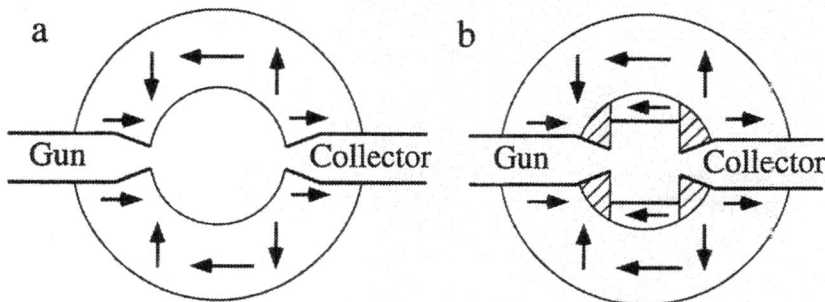

Figure 37. Magic spheres with electron guns and collectors. (a) Basic sphere, (b) Doubly augmented sphere.

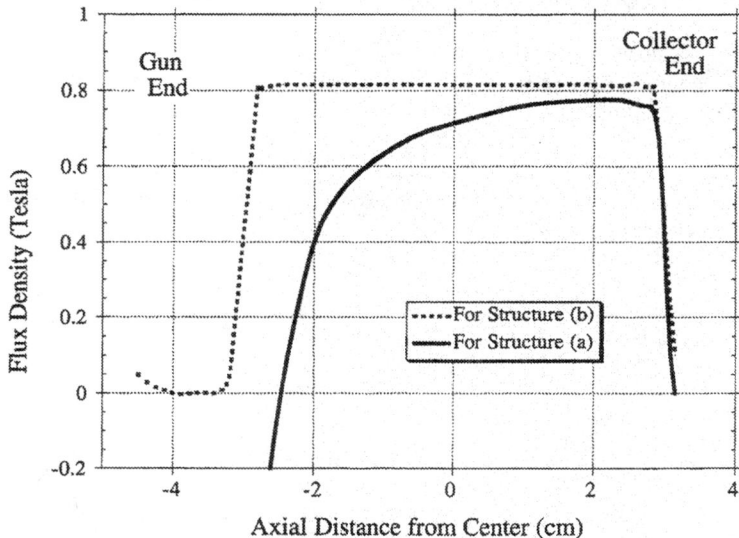

Figure 38. Field profiles in the working spaces of the structures of Fig. 37.

43

sphere. The resultant can be made to be zero anywhere within the spherical shell by the proper uniform addition. Figure 39a, for example, shows how the required magnetization can be made to be zero at the poles. Then, since no magnetization is required at the poles, the material there can be removed with impunity thereby affording access. Figure 39b shows the provision of similar distortion-free access in the equatorial plane.

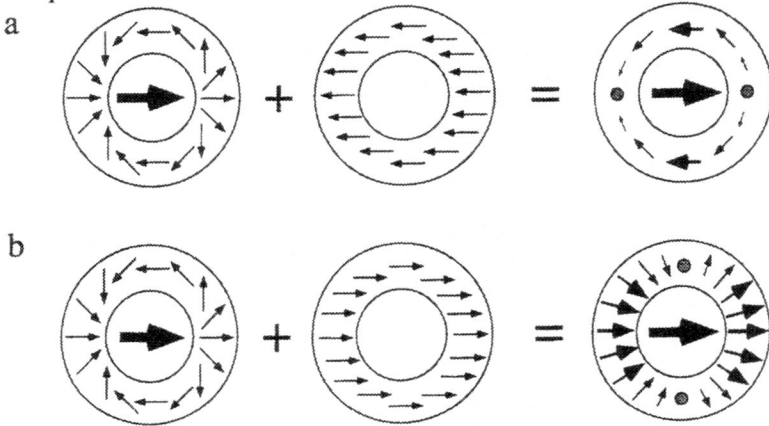

Figure 39. Production of distortion-free access for magic spheres; (a) for polar access, (b) for equatorial access.

This method has the disadvantage that the magnitudes of the largest magnetizations required in the altered sphere are twice that of the original sphere. If the original sphere already employed the highest magnetizations available, the altered sphere could not be built. One can circumvent this difficulty by increasing the outer radius in accordance with the formula of equation 35 and Section 6. Of course, this would entail a considerable increase in mass but would be worthwhile where field integrity is more important than light weight. Figure 40 shows the improvement of field profile when this procedure is used on magic spheres with large access ports.

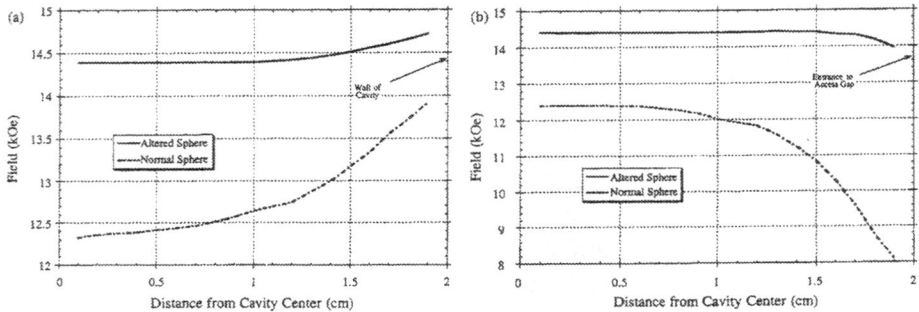

Figure 40. (a) Polar and (b)Equatorial axis field profiles for normal and altered spheres with 22.5° equatorial gaps.

44

To effect the same results without a large increase in mass, we note that much of the access-compensated sphere employs magnets of less remanence than that available. If we were to use the maximum available remanence everywhere, we could pare off material in sectors where less than maximum remanence is called for so that the product $B_r \ln (r_o / r_i)$ remains the same everywhere. If this is done, the structures of Fig. 41c and 41e result for polar and equatorial access, respectively.

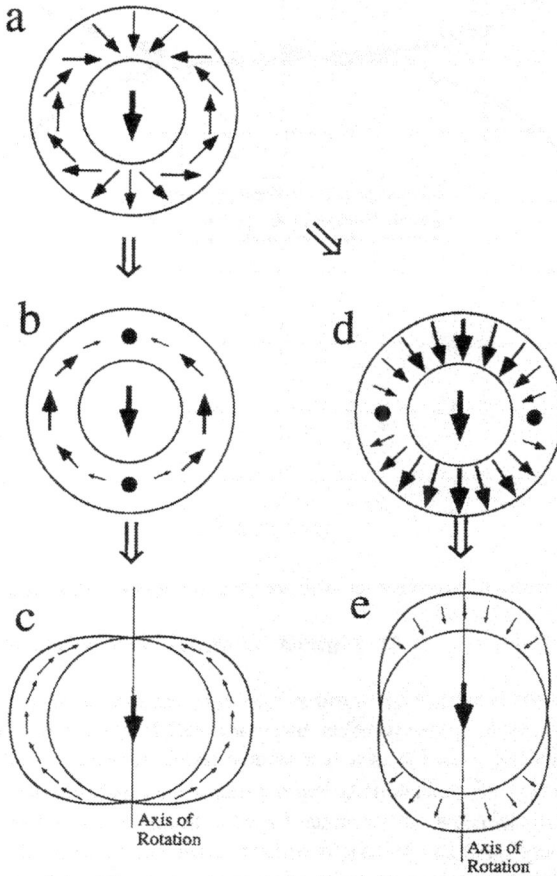

Figure 41. Geometric conversion of parametrically distortion-free access spheres to reduce mass.

For the polar access, the mass is still larger than for an uncompensated sphere but not as large as for one compensated geometrically. For the equatorial access, the geometrically compensated sphere is actually less massive than the uncompensated sphere by about 5%.

Geometric compensation is strictly valid only for the field at the center of the sphere and hence the "bumps" in the axial field profile of Fig. 42. These, however, are a small price to pay for maintenance of the full field near the ends of the profile. In any case, the bumps are tolerable for most beam focusing purposes and may be reduced by parametric means when greater precision is needed,

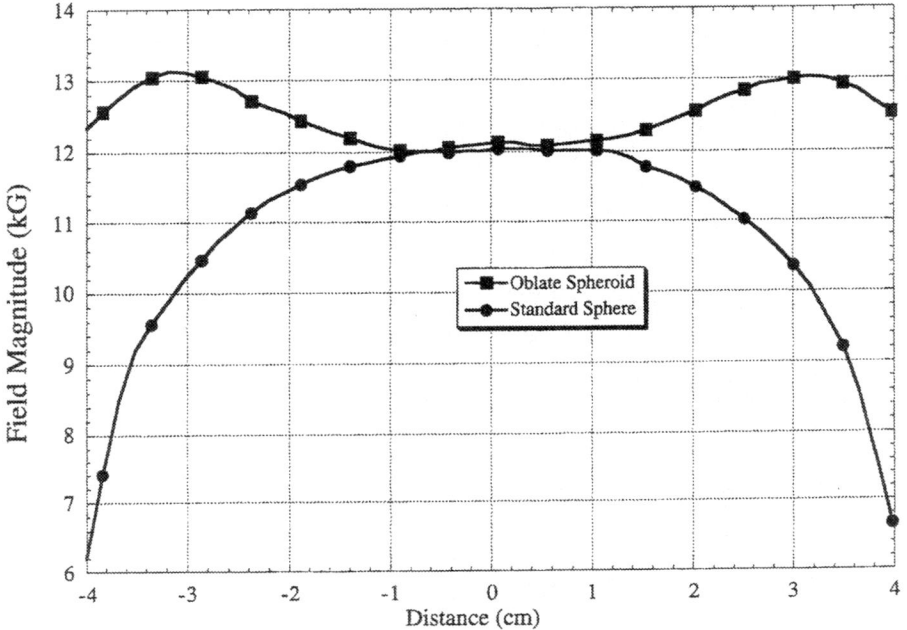

Figure 42. Field profiles of standard sphere geometrically converted for distortion-free access at the poles.

7. Structural Simplification with Magnetic Mirrors

All of these dipolar structures have anti-mirror symmetry in the equatorial plane perpendicular to their polar axes. For this reason, any of them can be cut at the equatorial plane and have either half placed on a high-permeability, high-saturation slab of material such as iron or permalloy to produce the same field in the interior that is produced in the cavity of an entire structure.[30] This works because the anti-mirror image of the half-structure formed in the slab produces the same field in the hemispherical cavity that is produced by the missing half. Such a procedure affords a non-trivial structural simplification since only slightly more than half as many different magnet segments of complex form need be manufactured for the semi-structure. The working volume is reduced by half, but if the full volume of the working space of the whole structure is needed, it may pay to use a larger half-structure to furnish it. Although more magnetic material would then be used, this drawback may be compensated by the increased simplicity afforded by use of fewer pieces.

An igloo-like structure made of half a sphere and a slab of iron is pictured in Fig. 43. Such a source made of Nd-Fe-B could provide a working flux density of 2 T in a 2.5 cm cavity with an outer diameter

46

of only 9 cm and a weight of 3 kg of permanent magnet material of 14 T remanence. This is an easily portable source and is potentially a very convenient general-use laboratory magnet. It also has the advantage of an easily accessible interior as lead holes can be drilled through the expendable iron slab rather than through expensive magnet material as in the sphere. This advantage affords greater flexibility with regard to modes of operation as iron slabs with different numbers, positions and sizes of holes would be easily interchangeable. Another advantage of the igloo is that transit through the iron is effected without encounter of field reversals. This is important for some electron-beam and Faraday rotator applications.

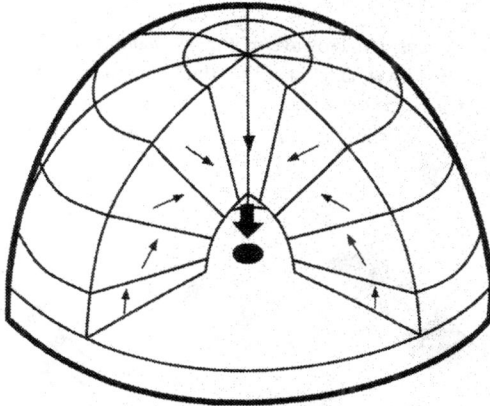

Figure 43. Section of a hemisphere of a spherical field source like that of Fig. 30 rests on an iron slab which takes the place of the missing hemisphere.

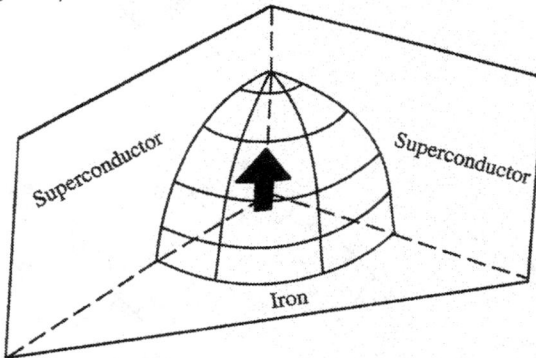

Figure 44. An eighth-sphere structure that produces the same field as a complete sphere with magnetic mirrors (superconductors), and anti-mirrors (iron). For simplicity, only the exterior and direction of polarity are shown.

If room-temperature superconductors of sufficiently high lower critical field were ever to become available, it would be practical to further simplify such structures by providing actual rather than anti-mirror images in an axial plane.[30] With various combinations of perfectly diamagnetic ($B=0$)

47

superconducting mirrors combined with perfectly paramagnetic ($B_0=0$) iron anti-mirrors, many configurations would be possible. The octant structure is an especially impressive example as it reduces the practical minimum of 120 segments needed to approximate a spherical source to only 15, as shown in Fig. 44.

8. Periodic Permanent Magnet Stacks (PPM's)

8.1 Travelling Wave Tubes

Travelling wave tubes are sources of microwave/millimeter wave radiation that employ an electron beam which is focussed and guided by an axial magnetic field. Sometimes the field is furnished by an electric solenoid but solenoids tend to be bulky and depend on cumbersome power supplies. In military devices, especially airborne and ballistic ones, mass and bulk must be minimized. To obtain the

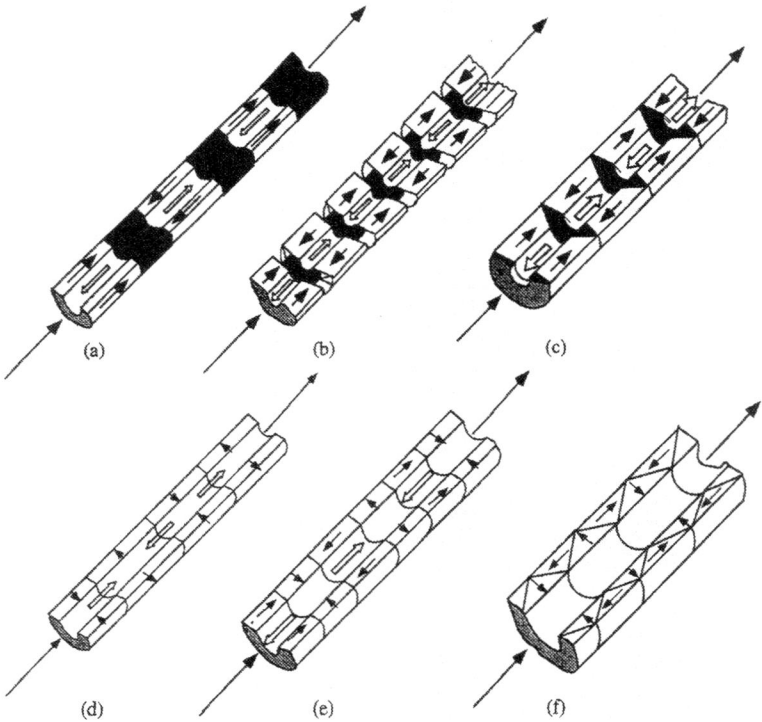

Figure 45. a. Simple PPM stack.   b. Stack with indented pole pieces. c. Stack with tapered pole-pieces. d. Radially oriented magnets. e. Hybrid PPM stacks. f. Version of e with triangular magnet cross sections. Small arrows show magnet orientations. Large open arrows show the working fields in internal cavities. The long arrows at the ends show electron beams. Darkly shaded areas are iron pole-pieces.

necessary high field intensities in small-bore cylindrical structures, the fields are usually made periodic and are supplied by appropriate permanent magnet stacks.

The simplest configuration is of the form shown in Fig. 45a where axially oriented magnets supply magnetic flux to the interior by way of interspersed iron pole pieces. Unfortunately, most of the the the flux leaks to the exterior rather than passing through the bore where it is wanted. The pole pieces can be indented or tapered at their external boundaries to increase the reluctance of the external flux paths to reduce the flux there as in figures 45b and 45c, respectively. This is what is usually done,[31] however the rigid permanent magnets make alternatives to this arrangement possible. An obvious stratagem is to use permanent magnets that are magnetized radially and are stacked to alternate between inward and outward orientations as in Fig. 45d. Such a stack is compared with the conventional array with regard to field and size in Fig. 46. The comparison shows a significant advantage in favor of the radial magnet

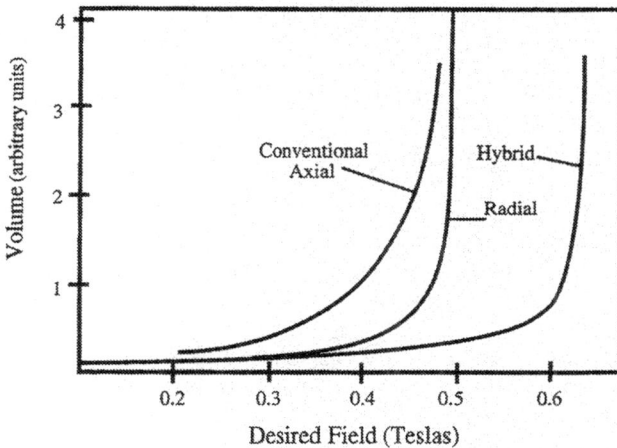

Figure 46. Comparison of bulks of structures of Fig. 45.

stack, especially in the technologically interesting region between 0.25 to 0.50 T where a conventional stack must be three times as massive to attain equal field with the same bore and period. An even better structure may be achievable by the use of the principle employed in the transverse circular cylindrical magnets already discussed. There large fields are obtained through gradual linear variation of the magnetization orientation with the coordinate angle ϕ. If the change in orientation of the magnetization with distance x along the axis of the radial magnet stack is made to be more gradual rather than in abrupt 180° jumps, the much improved field-bulk curve of Fig. 46 results. Here the size advantage over the conventional stack at the fields of interest is of an order of magnitude and more, while at the higher fields it rises by two orders of magnitude. However, gradual variation of magnetization is not technologically feasible. Fortunately, as in the case of the rings, rather coarse orientation increments of 90° between adjacent segments, as in Fig. 45e, produce a field ninety-four percent of that produced by the continuous variation. It is this insensitivity that changes the concept of gradual variation of orientation from an interesting speculative concept to a powerful technological tool.

Although the hybrid TWT magnets have impressive mass and field advantages over the conventional designs, they are more difficult to build for two reasons. First, it is hard to manufacture small-bore radially oriented ring magnets because of the difficulty of guiding sufficient flux through their bores to radially align and magnetize the high coercivity materials. Secondly, because there are no iron pole pieces through which small amounts of perturbing flux can be conveniently led to compensate for small field irregularities incurred in manufacture and assembly, all-magnet stacks are much harder to balance.

A simple and very effective alternative is to make pole pieces of triangular toroidal cross section as in Fig. 45c.[31,32] This increases the magnetic permeance of the bore relative to that of the exterior so that more flux flows through the former. It also increases the average effective length of the axially oriented magnets, thereby increasing the magnetomotive force that drives the magnetic circuits. Both effects increase the field amplitude in the bore.

While it does not afford the spectacular mass and bulk advantages of the hybrid stack, the triangular poled array is still smaller by a factor of two than that of a conventional stack with indented poles at a bore field amplitude of 0.5 T. Moreover, it has none of the manufacturing and balancing difficulties associated with the hybrid stack. It is probably the best all-around practical alternative at present.

8.2 Wigglers and Undulators

Periodic transverse field sources are called wigglers and are used to provide lateral oscillatory motion to electron beams in free-electron lasers. It is this lateral acceleration that causes the electron to radiate. If the product of the flux density in T and the wiggler period in centimeters is of the order of one or less, the beam will radiate coherently and laser action results. When this occurs, the wiggler is called an undulator.[13,33]

The resonance condition is easily derived. If Gaussian units are used the cyclotron frequency of an electron is given by

$$f = eB / 2\pi mc$$

and for the usual relativistic beam the frequency of the field oscillation is

$$f_c = c / l$$

where l is the period of the wiggler. For resonance to occur these must be equal:

$$f_c = f_w$$

$$Bl = 2\pi mc^2 / e = 10,700 \text{ G-cm}$$

or if B is in T

$$Bl = 1.07 \approx 1 \text{ T-cm}$$

an easy to remember condition.

As usual, when magnetic fields with a rapid spatial variation are required, permanent magnets offer the best solution. The simplest permanent magnet wiggler is just a series of bar magnets arranged as in Fig. 47a. More efficient would be planar arrays of magnets of alternately longitudinal and transverse orientation (see Fig. 47b). The principle used here is that of the hybrid TWT structure and, as in that structure, a high field-to-mass ratio results. But like the TWT hybrid, it is somewhat difficult to adjust because of its lack of iron pole pieces to which one can attach compensating shims. For this reason, the less bulk-efficient but more tractable arrangement of Fig. 48 is often used.[13] It can produce very high

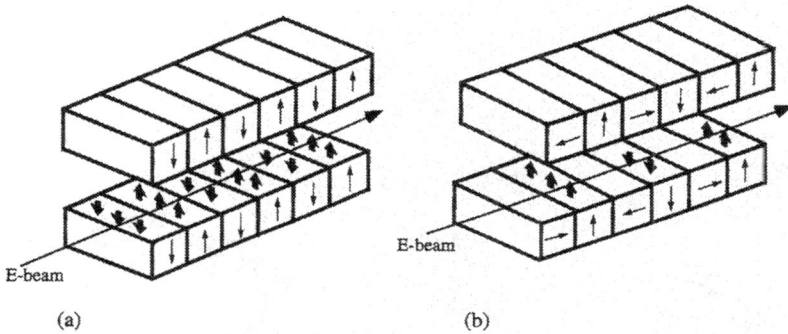

(a)   (b)

Figure 47. Simple wiggler structure, (a). High field structure with 180° change in orientation of successive magnets, (b). Hybrid structure with 90° change in orientation of successive magnets.

fields, is simple, relatively easy to adjust and probably the most practical choice where minimum mass and bulk is not a consideration. It has the advantage over the similar arrangement of Fig. 47a in that its indented pole pieces allow less flux leakage.

Axial arrays of cross-sectional slices of any of the cylindrical transverse structures already discussed have the advantages of high-field capability and minimal flux leakage to the structural exterior, but they also suffer from relative complexity and are not readily adjusted by shims.

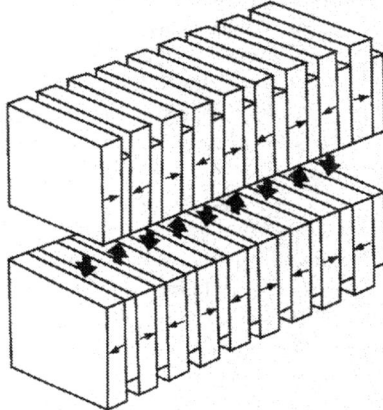

Figure 48. Simple wiggler stack with alternately oriented longitudinal magnet interspersed with iron pole pieces.

8.3 Twister Structures

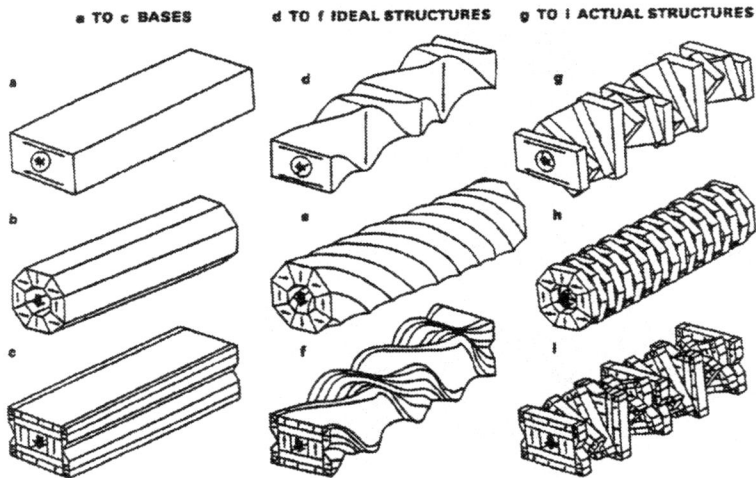

Figure 49. The three twister structures (**d-f**) with their respective transverse cylindrical bases (**a-c**) and their practical embodiments (**g-i**).

The radiation from wigglers and undulators is polarized in the plane of the oscillatory motion of the electron beam. Sometimes it is desirable that the radiation be circularly polarized in which case a different magnetic array is needed to impart the necessary helical motion to the electron beam. This can be accomplished by employment of either a bifilar helical electric solenoid or by stacking a series of wiggler elements so that successive elements differ in orientation by a small angle φ. This results in a field that is constant in magnitude but whose orientation changes continuously with progression along the beam axis. Figure 49 shows such arrays together with those with untwisted cylinders from which they are derived and the stacks of finite segments with which they are approximated in practice. The structures shown in Figs. 49b and 49c will be recognized as ones having already been discussed in cylindrical form. The structure in Fig. 49a is a simple, transversely magnetized bar magnet with a hole bored along its axes to accommodate the electron beam. The magnetic field in the hole is generated by the dipolar distribution on the inner surface that arises from the constant magnetization of the material. This field is slightly reduced by the presence of the more remote poles that the magnetization forms on the rectangular ends. The field is also progressively reduced with rate of twist and coarseness of twist increments. Fortunately, the effect of the latter is small for reasonable increments. For example, the 36° change between successive segments used in a prototype results in a field diminution of the order of only one percent. Figures 50 and 51 compare the properties of the three structures with regard to parameters of interest. Although not quite so desirable as the structure of Fig. 49b, the simple bar magnet of Fig. 49a was chosen for a prototype design for a high powered radar source which is still far superior to the solenoid it will replace and less expensive than the other structures. The only advantage

52

of the configuration in Fig. 49c is that the iron equipotential it forms tends to smooth aberrations in the material and structure to give a more uniform field. The twister design was chosen over the wiggler because circularly polarized radiation affords better discrimination on some military targets. Also, in a twister the magnetic field is at maximum throughout its length while a wiggler produces an average field of approximately 0.707 maximum.

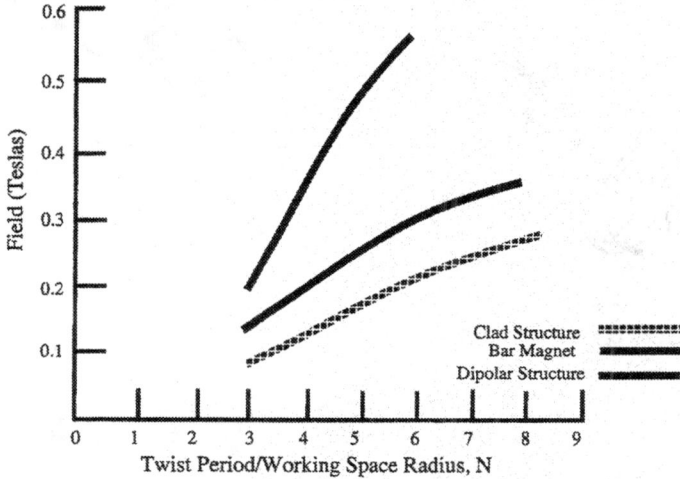

Figure 50. Maximum field attainable in three twister structures as a function of twist period to working space ratio.

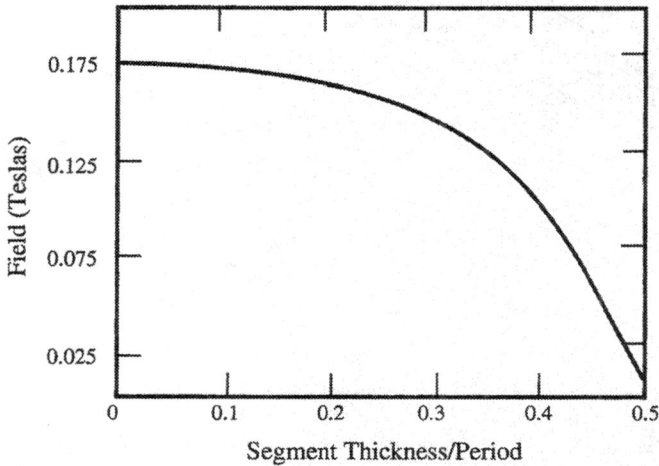

Figure 51. Maximum field attainable as a function of segment thickness/period ratio for the octagonal structure of Fig. 49b.

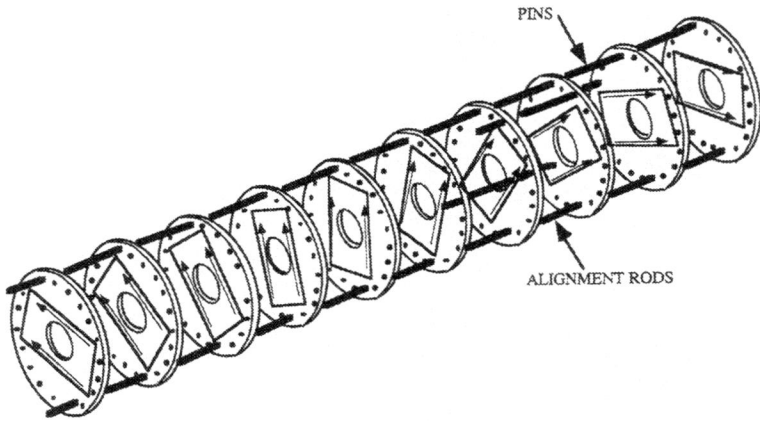

Figure 52. Exploded view of a twister prototype.

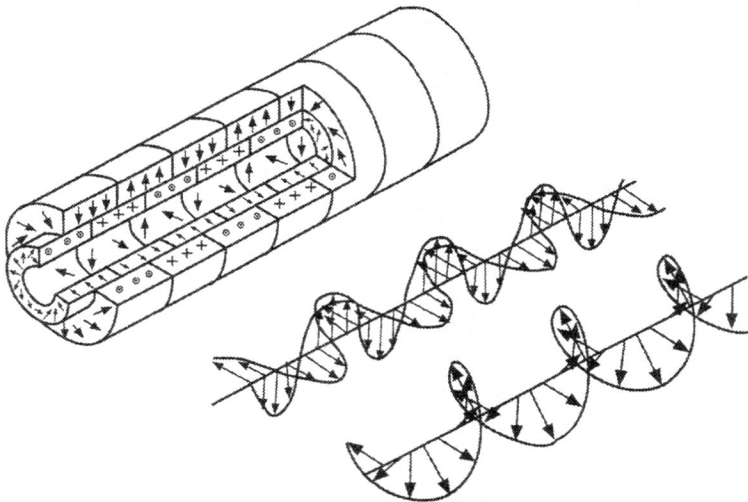

Figure 53. Addition of fields of two nested twisters to form a wiggler field.

The prototype was built and found to deliver the theoretical field of 0.13 T compared to the 0.06 T delivered by the presently employed solenoid with current of 200 amperes. Structural details are shown in Fig. 52. The magnet slices are mounted in brass discs for mechanical protection. Rods through holes near the periphery of the brass mounts offer the proper alignment and provide flexibility in twist rate so

54

that pitch can be adjusted or even varied with progression along the axis. The field strength and distribution agrees with the design specifications to about one percent.

A wiggler made up of field adjustable nested rings discussed in the previous section affords an intriguing possibility.[19] If the inner stack of such a structure is rotated by 90° and displaced one quarter period along the axis, its field will add vectorially with that of the outer ring to form the helical pattern of a twister (see Fig. 53). Thus, the adjustable ring wiggler is not only quantitatively adjustable, but can be operated as a twister of the same frequency as well.

A recent development in free-electron laser technology is that of the CHIRON-type wiggler. It was invented by Freund and Jackson at NRL to produce higher periodic field amplitudes in short-period wigglers by employment of the high saturation magnetization of iron alloys. Figure 54 shows a simplified version of the design actually made.

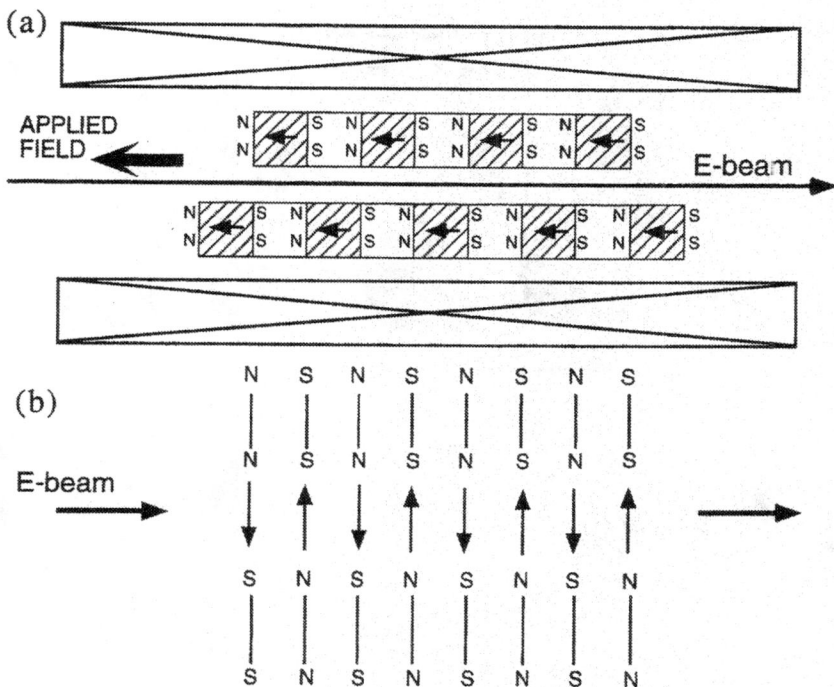

Figure 54 Simple CHIRON type wiggler. ▨ is iron. (a) iron vanes in applied field, (b) arrows show resulting polar distribution

The iron vanes in the drawing are saturated by a strong externally applied field so that the pole distribution shown in the lower part of the drawing is formed. Since the pole density in the polar sheets is proportional to the magnetization which forms them and the 2.4T saturation value of the strongest iron alloys is twice the 1.2 T of a high remanence permanent magnet, the charge sheets of the iron will be twice as dense. The vanes are staggered on opposite sides of the gap so that opposite polarities face each other across it. In the NRL wiggler, the structure is cylindrical and the gap is of annular shape, so that a hollow electron beam travels in the axial direction and vibrates azimuthally in the process. Hence, the name CHIRON wiggler for Coaxial Hybrid-Iron. At the Physical Sciences Directorate of the Army Research Laboratory, it was recognized that a simple twister structure could be made by employment of the same principle. A hollow iron pipe has a helical slot cut through its wall so that the half-period, $\lambda/2$, of the resulting helical bar of rectangular cross section is equal to its axial thickness τ. When the bar is placed in an axially applied saturating field, magnetic poles with surface pole density form on left and right surfaces as shown in Fig. 55. The result is two parallel oppositely poled ribbons separated

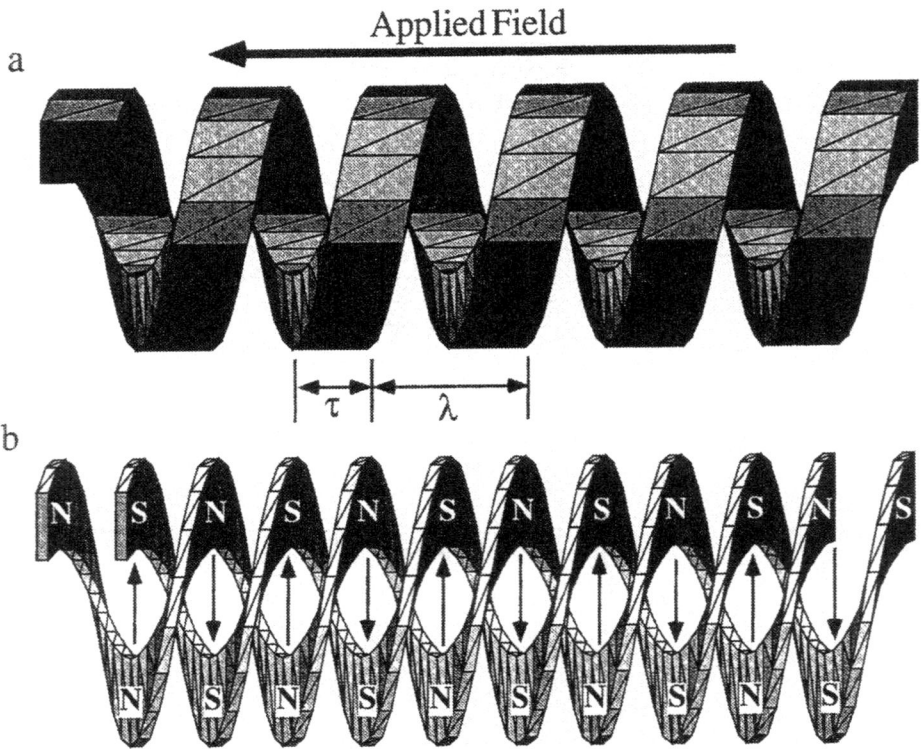

Figure 55. (a) Iron helix in applied saturating field, (b) resulting polar distribution.

by distance τ so that opposite poles are at the ends of any inner diameter. The on-axis field generated by this distribution is diametric in direction and constant in value and undergoes a rotation of 2π with an advancement of $\lambda=2\tau$ along the axis. Such a field is that of a helical free-electron laser or twister.

Figure 56 shows the field produced for two different twist-to-radius ratios λ/r_i where r_i is the inner radius of the structure as a function of twister mass per unit length. The figure also compares this structure with a more conventional twister designed for NRL. The field-to-mass advantage of the former is obvious. Figure 57 compares the two twisters with regard to maximum field attainable at a range of twist values and again the CHIRON-type enjoys a substantial advantage except at the lowest twist ratios although it is marginally better even there.

The CHIRON-type structure can also be formed from a permanent magnet but then the field would be reduced by approximately half because of the relative magnetization ratio of approximately 2:1 of the iron to permanent magnet. Figure 58 shows how two twister structures of opposite polarity can be threaded together to form pole layers of twice the density of a single unit and hence twice the field. This would raise the field to approximately that of the iron CHIRON or even slightly more if newer permanent magnet materials of higher remanence are used. This procedure would eliminate the need for a massive saturating coil with its power supply and cooling system, but would not afford an adjustable field as does the electromagnet.

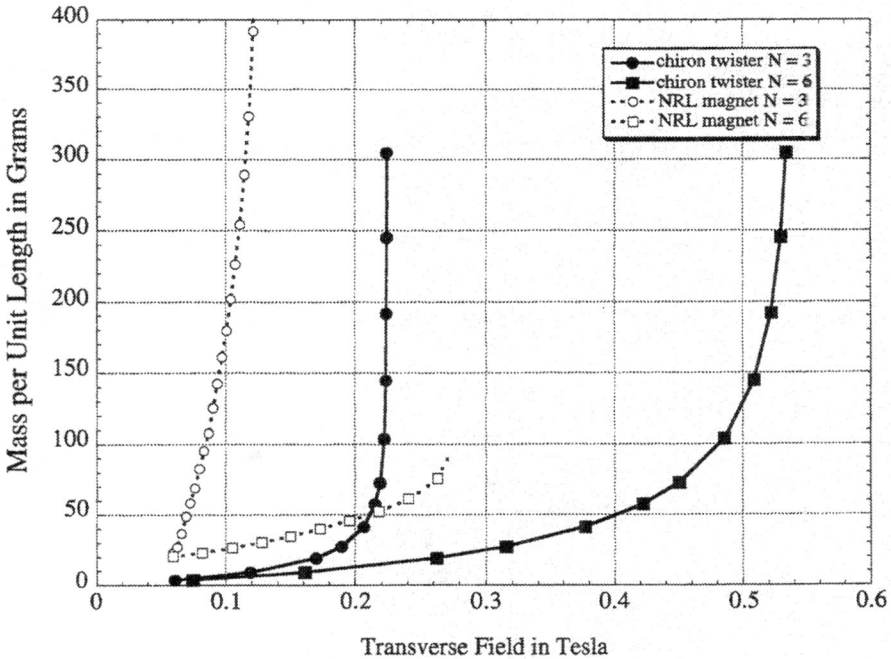

Figure 56. Comparison of fields in CHIRON versus NRL twister magnets. N is the ratio of twist to gap radius.

57

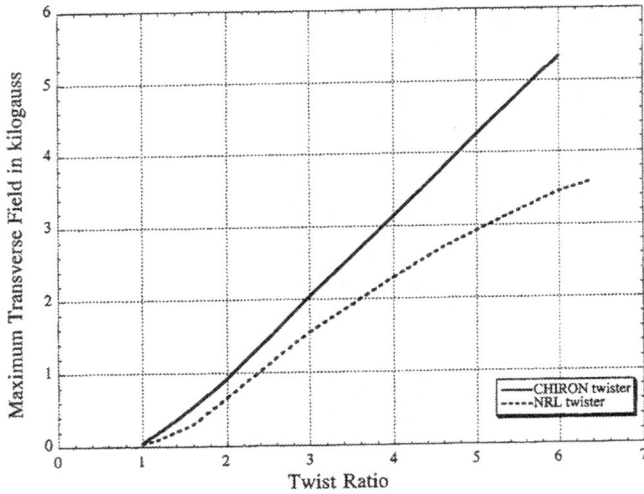

Figure 57. Comparison of maximum transverse field versus twist ratio.

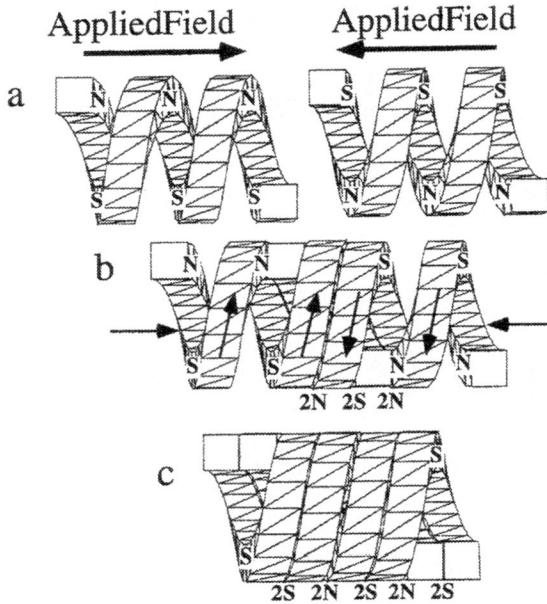

Figure 58. Double helix twister.

9. Conclusions

While much progress has been made in rare earth permanent magnet technology over the past two decades, the demand for improvements seems to be perennial.[1,2] As projected devices become more complex, the technological demands become more stringent. There is considerable room for improvement in the following areas.

(1) Better temperature performance of temperature-compensated magnets. This includes temperature coefficients of the magnets' energy products and their extension to broader temperature ranges. Also important is further improvement of remanence and loop squareness. These developments are absolutely essential for the attainment of higher fields in smaller structures operating at greater power densities in TWT's projected for use in future lightweight, airborne devices. Raising the coercivities and squaring of the loops of the currently best Sm_2Co_{17} temperature-compensated magnets might be a solution. Compensation of the Nd-Fe-B compounds might be an even better one as their energy products are generally much higher than those of the Sm_2Co_{17}s but suffer from low Curie temperatures.

(2) Improved magnet alignment in novel magnet geometries by getting sufficient flux where it is needed during the fabrication and magnetization processes. Sometimes this is very difficult in the case of conical shell magnets aligned normal to the cone surfaces, which are needed for the cladding of pole pieces in horseshoe-type magnetic circuits. Also difficult, for the same reason, is the radial magnetization of toroids with small holes and broad annuli.

(3) Improvement of fabrication and assembly techniques through the development and use of better cements and dies.

(4) A better variety of square-looped materials with low energy products is needed for widespread implementation of parametric field shaping. This might be done by the development of very high coercivity ferrites, or the dilution of oriented materials in a metal or epoxy matrix.

Investigation of deposition of REPM materials in the form of films on surfaces of different geometries should continue to be pursued.[34] Energy products must be raised, hysteresis loops squared, and mechanical and thermal ruggedness assured. Especially promising is the possibility of deposition of films of in-surface and normal-to-surface orientations upon each other. Success in such a venture could lead to a whole generation of small devices that are unattainable at present, e.g., miniature field sources for microelectronics, micromotors and miniature wigglers.

In summary, the REPMs are very much more than just very powerful Alnicos. They afford not only the possibility of doing conventional tasks better, but offer a cornucopia of design options that were impossible before their advent. As mindsets acquired in the course of design experience with conventional magnets are discarded, the technological riches offered by rare earth permanent magnets can be fully and speedily realized.

59

10. References

1. F. Rothwarf, L.J. Jasper and H. A. Leupold, Proceedings of the Third International Workshop on Rare Earth-Cobalt Permanent Magnets and Their Applications, University of California, San Diego, June 27-30 (1978), p. 255; Proceedings book by University of Dayton, School of Engineering, Dayton, Ohio, USA, 45469

2. A. Tauber, H. A. Leupold and F. Rothwarf, Proceedings of the Eighth International Workshop on Rare Earth-Cobalt Permanent Magnets and Their Applications, Dayton, Ohio USA, May 5-9 (1985), p. 103

3. H. A. Leupold, Proceedings of the Fifth International Workshop on Rare Earth-Cobalt Permanent Magnets and Their Applications, Roanoke, Virginia USA, June 7-10 (1981), p. 270

4. Parker and Studders, "*Permanent Magnets and Their Applications,*" John Wiley and Sons, Inc., New York, London (1962)

5. H. C. Roters, "*Electromagnetic Devices,*" John Wiley and Sons, Inc., New York, and Chapman and Hall Ltd., London

6. H. A. Leupold, F. Rothwarf, C. G. Campagnoulo, H. Lee and J. E. Fine, "*Magnetic Circuit Design Studies for an Inductive Sensor,*" U.S. Army Technical Report, TR-ECOM-4158, October (1973), Fort Monmouth, New Jersey USA, 07703

7. W. Neugebauer and E. M. Branch, "*Applications of Cobalt-Samarium Magnets to Microwave Tubes,*" Technical Report, Microwave Tube Operations, General Electric, 15 March (1972), Schenectady, New York USA

8. J. P. Clarke and H. A. Leupold, IEEE Trans. Magn. **MAG-22**, No. 5, p.1063 (1986)

9. H. A. Leupold and E. Potenziani II, IEEE Trans. Magn. **MAG-22**, No. 5, p.1078 (1986)

10. H. A. Leupold, E. Potenziani II and J. P. Clarke, "*Shaping of Cylindrically Symmetric Magnetic Fields with Permanent Magnets,*" U.S. Army Technical Report, DELET-TR-84-12, December (1984), Fort Monmouth, New Jersey USA, 07703

11. J. P. Clarke, E. Potenziani II and H. A. Leupold, J. Appl. Phys. **61** (8), p. 3468(1987)

12. K. Halbach, Proceedings of the Fifth International Workshop on Rare Earth-Cobalt Permanent Magnets and Their Applications, Roanoke, Virginia USA, June 7-10 (1981), p. 73

13. K. Halbach, Proceedings of the Eighth International Workshop on Rare Earth-Cobalt Permanent Magnets and Their Applications, Dayton, Ohio USA, May 5-9 (1985), p. 103

14. K. Halbach, Nucl. Instr. and Meth. **169**, p. 1 (1980)

15. K. Halbach, Nucl. Instr. and Meth. **187**, p. 109 (1981)

16. R.F. Holzinger, Proceedings of The Sixth International Workshop on Rare Earth Permanent Magnets and Their Applications, Editor J. Fidler, Tech U. of Vienna, p. 147 (1982)

17. H. Zijlstra Philips J Research **40** p. 259 (1985)

18. A. B. C. Morcos, H. A. Leupold and E. Potenziani II, IEEE Trans. Magn. **MAG-22**, No. 5, p. 1066 (1986)

19. H. A. Leupold, E. Potenziani II and M. G. Abele, J. Appl. Phys **64**, No. 10, p. 5994 (1988)

20. O. Cugat, R. Byrne, J. McCaulay and J. M. D. Coey, TBP Review of Scientific Instruments, 1994.

21. J. M. D. Coey and O. Cugat, Proceedings of International Workshop on Rare Earth Permanent Magnets and Their Applications, September (1995)

22. H. A. Leupold, E. Potenziani II, J. P. Clarke and D. Basarab, Material Research Society Symposium, **96**, p279 (1987)

23. M. G. Abele and H. A. Leupold, J. Appl. Phys. **64**, No 10, p. 5988 (1988)

24. M. G. Abele, R. Chandra, H. Rusinek, H. A. Leupold and E. Potenziani II, IEEE Trans. Magn. **MAG-25**, No. 5, p. 3904 (1989)

25. M. G. Abele and H. Rusinek, J. Appl. Phys. **67**, No. 9, p. 4644 (1990)

26. M. G. Abele, Technical Report, New York University School of Medicine, TR-20, 1 March 1989

27. M. G. Abele "Structures of Permanent Magnets," John Wiley & Sons Inc., New York (1993)

28. M. G. Abele, Technical Report, New York University School of Medicine, TR-15, 1 March 1989

29. H. A. Leupold, E. Potenziani II and A. S. Tilak, IEEE Trans. Magn. **MAG-28**, p. 3045 (1992)

30. H. A. Leupold and E. Potenziani II, J. Appl. Phys. **63**, 8IIB, p. 3487 (1988)

31. H. A. Leupold, E. Potenziani II and A. Tauber, J. Appl. Phys. **67**, No. 9, p. 4656 (1990)

32. H. A. Leupold and J. P. Clarke, IEEE Trans. on EDS, Part II, Vacuum Electron Devices, **ED-34**, No. 8, p. 1868 (1987)

33. Thomas C. Marshall, "Free Electron Lasers," Macmillan Publishing Co., New York, NY (1985)

34. F. J. Cadieu, T. D. Cheung, L. Wickramasekara and S. H. Aly, J. Appl. Phys. **55**, p. 2622 (1984)

11. Bibliography

(1)   Parker and Studders,*"Permanent Magnets and Their Application,"* John Wiley and Sons, Inc., New York, London (1962).
Introductory work to pre-REPM technology. It discusses some elementary theory, manufacturing procedures, magnet properties, design techniques and applications. It contains a number of useful graphs, tables and charts summarizing material properties and magnetic circuits.

(2)   Rolin J. Parker, *"Advances in Permanent Magnetism."*
Partly an update of (1), but with the inclusion of REPMs, more circuit theory, and an expanded, modernized section on magnetic instrumentation and measurement.

(3)   Herbert C. Roters,*"Electromagnetic Devices,"* John Wiley and Sons, Inc., New York; Chapman and Hall, Limited, London (1941).
Covers most practical aspects of magnet design in the pre-computer, pre-REPM era. It includes the most extensive treatment known to the authors of the estimation of permeance method of magnet design. Has copious illustrative problems as well as drill problems at the ends of chapters.

(4)   Richard M. Bozorth,*"Ferromagnetism",* IEEE Press, Piscataway, New Jersey.
A reissued old classic. Still very useful for its compendia of passive magnetic material properties and general magnetic lore.

(5)   Manlio G. Abele, *"Structures of Permanent Magnets,"* John Wiley and Sons, Inc. (1993).
A systematic and mathematically rigorous study of the design of cylindrical magnetic shells that confine uniform transverse fields to their inner chambers. The cylinders are of arbitrary cross-sectional shape. A near-indispensable work for efficient REPM design.

(6)   H. Zjilstra, Philips J. Res., 40, p. 259 (1985).
A useful summary of specific transverse field structures for cylindrical cavities. The consequences of finite length are discussed in detail. There is some discussion of spherical structures as well.

www.ingramcontent.com/pod-product-compliance
Lightning Source LLC
Chambersburg PA
CBHW061839220326

41599CB00027B/5347